Advances in Maritime Logistics and Supply Chain Systems

Advances in Maritime Logistics and Supply Chain Systems

editors

Ek Peng Chew
Loo Hay Lee
Loon Ching Tang

National University of Singapore, Singapore

World Scientific

NEW JERSEY · LONDON · SINGAPORE · BEIJING · SHANGHAI · HONG KONG · TAIPEI · CHENNAI

Published by

World Scientific Publishing Co. Pte. Ltd.

5 Toh Tuck Link, Singapore 596224

USA office: 27 Warren Street, Suite 401-402, Hackensack, NJ 07601

UK office: 57 Shelton Street, Covent Garden, London WC2H 9HE

British Library Cataloguing-in-Publication Data
A catalogue record for this book is available from the British Library.

ADVANCES IN MARITIME LOGISTICS AND SUPPLY CHAIN SYSTEMS

ISBN-13 978-981-4329-85-9
ISBN-10 981-4329-85-1

Typeset by Stallion Press
Email: enquiries@stallionpress.com

Printed in Singapore by World Scientific Printers.

CONTENTS

PREFACE

Over the recent years, maritime logistics and supply chains have witnessed tremendous growth rates around the world, notwithstanding the recent economic downturn. Maritime transportation accounts for the majority of international trade and it has become a vital factor for the economic health of many nations. In emrging economies, more new ports have also been developed to tap into the global maritime logistics network. The global landscape of the martime industy is changing rapidly and this has generated many issues which are worthy of more in-depth research. In particular, topics related to maritime logistics and supply chains have been drawing immense attention of both academia and industry.

The objective of this book is to reflect the recent developments in maritime logistics and supply chains, and to examine some research issues concerned with quantitative analysis on port competitiveness and decision support for maritime logistics and supply chain systems. Twelve papers have been selected for publication after a thorough peer review. The papers are categorized into two main areas: regional developments and performance analysis; and, ports and liners operations.

Regional Developments and Performance Analysis

The first paper by Xue-Jing Yang, Joyce M.W. Low and Loon Ching Tang, tracks maritime trade evolution in Asia from the thirteenth centuries to the post-World War II, followed by an examination on the contemporary development of some major Asia ports. Some factors affecting port competition and development are identified and reviewed. Their study concludes with their beliefs that maritime trade industry in Asia is promising and positive future economy trends will continue despite recent concerns over rising oil prices.

The paper by L. H. Lee, E. P. Chew, L. Zhen, C. C. Gan, and J. Shao presents the recent development in Maritime Logistics during the recent economic crisis. The recent development of container shipping industry has shown the following trends: (1) The size of the largest container vessel and the average vessel size are both increasing. (2) Transshipment handling has become more and more significant globally. (3) Global container terminal operators are increasing their market share. (4) Liner companies are adopting more rigorous measures to reduce cost and stabilize freight rate.

Hong Kong port had been the world's busiest container port during the 1990s and early 2000s. However, in recent years, its growth slowed down due to rising competition from mainland ports. Abraham Zhang and George Q. Huang perform some scenario analysis for Hong Kong port development under changing business environment, so as to understand the relationships between business environment factors and potential relocation trends by using a mixed integer programming model.

A study on port benchmarking in Asia-Pacific region is performed by Ek Peng Chew, Loo Hay Lee, Jianlin Jiang and Chee Chun Gan. Some models for port competitive analysis are proposed from three perspectives: port efficiency, port connectivity, and the impact of various factors on individual ports. The authors examine the three perspectives by presenting a model and a case study for each perspective. Data envelopment analysis technique, a port connectivity analysis framework, and a network flow model are proposed and employed to investigate the above three perspectives respectively.

Wayne K. Talley investigates the question: is port throughput a port output in port economic production and cost functions? His study disagrees on this statement. The author proposes the port throughput ratio — the ratio of cargo interchanged to the total time incurred in interchanging the cargo, for measuring the output of a port.

The paper by Vassilios K. Zagkas and Dimitrios V. Lyridis proposes a framework for modeling and benchmarking maritime clusters. The authors investigate the factors that contribute to the decisions of companies from key maritime sectors to be established in a specific area that evolves into a network of firms. Agent-based modeling technology is employed to simulate the networking process within maritime clusters and managing their life cycle. This study gives an insight of firm survival strategies within the cluster, optimum timing for new entrants in the cluster and overall cluster management.

The paper in this group by Min Chen, Wei Yan and Weijian Mi propose a performance evaluation strategy for dealers in the automotive supply chain. The performance evaluation strategy is developed from four dimensional criteria, i.e., the financial condition, customer satisfaction, internal processes and self-innovation. The analytic network process (ANP) technique is employed to analyze the surveyed data. By comparing with traditional performance evaluation strategies, their approach can eliminate such disadvantages as time delaying and benefit orientation.

Ports and Liners Operations

Container yard management is essential for the efficiency of terminal operations. A yard allocation strategy is proposed in the paper by Wei Yan, Junliang He, Daofang Chang. By using the objective programming, the proposed model is based on a rolling-horizon strategy, which aims at allocating export containers into yard. For solving the model, a hybrid algorithm by using heuristic rules and genetic algorithm is employed. A simulation model, which embeds the yard allocation model and algorithm, is also developed to evaluate the proposed system.

The paper by Bernd H. Kortschak examines the integration of railway links with other functions in a deep sea terminal. Currently, railway links are often operated separately which incurs additional costs when transferring containers and hinders the competitive strength of rail versus road links. An integrated AGV system is proposed to improve productivity and enable faster transshipment times.

In the paper by Simme Veldman, the author conducts a statistical analysis of economies of ship size. This study shows that economies of ship size, expressed as the elasticity of costs as a function of ship size, differ only slightly from those of ships up to Panamax. For avoiding too high cost for users, the increase in size has to be in balance with the combined increase in trade volumes and the number of port pairs between coast lines to be connected. The study draws a conclusion that the ongoing increase in ship size will continue.

In the paper by Qiang Meng, Tingsong Wang and Shahin Gelareh, a linearized approach is proposed for liner ship fleet planning under demand uncertainty. The authors develop a mixed integer nonlinear programming model for the problem; then the fuel consumption cost of a ship is approximated by a linear function with respect to its cruising speed. Hence a mixed integer linear programming model can be built to approximate the originally proposed nonlinear programming model. Numerical examples are performed to assess the linearized approach.

The paper by Harilaos N. Psaraftis and Christos A. Kontovas studies ship emissions, costs and their tradeoffs. The authors investigate various tradeoffs that may impact the cost-effectiveness of the logistical supply chain, and propose some models that can be used to evaluate these tradeoffs. Their study validates that speed reduction can lead to a lower fuel bill and lower emissions, even if the number of ships is increased to meet demand throughput. In addition, cleaner fuel at SECAs may result in a reverse cargo shift from sea to land that has the potential to produce more emissions on land than those saved at sea.

By using a simulation tool named by FORESIM, P.G. Zacharioudakis and D.V. Lyridis study the future tanker market freight levels in relation to current market fundamentals and future values of demand drivers. The authors follow a systems analysis seeking for internal and external parameters that affect market levels. By using the proposed methodology, decision makers can measure the behavior of future market as long as twelve months ahead with very encouraging results. The output information is potentially useful in all aspects of risk analysis and decision making in shipping markets.

Concluding Remarks

This book has greatly benefited from the cooperation among the authors, reviewers and editors. We would like to express our sincere thanks to them.

E.P. Chew

L.H. Lee

L.C. Tang

December 2010, Singapore

PART I

REGIONAL DEVELOPMENTS AND PERFORMANCE ANALYSIS

CHAPTER 1

MARITIME TRADE EVOLUTIONS AND PORT CITY DEVELOPMENTS IN ASIA

X.J. Yang, Joyce M.W. Low and Loon Ching Tang
Department of Industrial & Systems Engineering
National University of Singapore
1 Engineering Drive 2, Singapore 117576
isetlc@nus.edu.sg

Historically, almost all goods transported worldwide have been carried by sea with the current estimate stands at approximately 90 percent by volume and 70 percent by worth. Maritime industry is an important economic sector as it has a direct impact on the prosperity of a region and/or city. This chapter presents a review on maritime trade evolution in Asia from the thirteenth centuries to the post-World War II, followed by an examination on the contemporary development of some major Asia ports. From the extant port literature, a list of factors affecting port competition and development is identified and reviewed. The chapter concludes with brief discussions on fruits for thought, future trend, challenges and opportunities facing the Asia maritime trade industry.

1. Introduction

Maritime shipping represents the most ancient global transportation, holding an irreplaceable role in geography discovery, culture communication and economy development in history. As early as the fifteenth century, Chinese admiralty Zheng He had visited Indian ports and east Africa seven times through Southeast Asia. His famous voyages marked the beginning of the international maritime trade in Asia. Subsequently, international and global voyages in the following centuries had led the European admiralties to discover the North America and opened Africa continents, as well as, the principle of earth's as a spheroid planet. To some degrees, these illustrious maritime explorations had shaped the world's history. Particularly, international maritime trade has played a significant role in spreading civilization to many parts of the world and promoting communications that aid economic development.

In this modern era, maritime trade industry has gained unsurpassed vigor with more high-value cargo transported by seaborne shipping than before. Between 1990 and 2003, the value of maritime trade to the US economy ballooned from 434 billion dollars to 800 billion dollars. [DeGaspari, 2005] indicated that approximately 75 container ships cross the Pacific each week, making the trans-Pacific an economic engine between North America and Asia in 2005. In Europe, container port throughput has increased by some 58 percent to 69 percent, reaching 49.5 million Twenty foot equivalent unit (TEU) in 2001; and is expected to hit 53.0 million TEU in 2010. The transshipment market, despite volatility, has also grown by an average 9.7 percent per annum since 1995 in North Europe. In Asia, some major ports have replaced New York, Tokyo and Rotterdam ports as the top container ports in 2005 [Lee et al., 2008]. These ports have continued to grow and by 2008, Singapore, Hong Kong, Shanghai, Shenzhen and Busan have registered impressive container traffic of 29,918 thousand TEUs, 24,494 thousand TEUs, 27,980 thousand TEUs, 21,414 thousand TEUs and 13,453 thousand TEUs respectively.

Continual globalization of world economy, coupled with recent energy crisis, will propel maritime trade to a higher level in the near future. In Asia, maritime industry occupies a central position, not only for economies with large transshipment ports like Singapore and Hong Kong, but much more for China and Japan who have an increasing demand for oil. At present, China is the world's manufacturing center and ranked as the third largest economy with huge domestic market. This newly industrialized giant economy represents an exploding demand for oil and the Malacca Straits is the only major way for China to receive oil from Middle East and India currently. [Chua et al., 2000] substantiated the claim in noting that this trade route delivered over 100,000 oil and cargo vessels each year, 3.23 million barrels of crude oil each day. In addition, the Malacca Straits is also an important shipping route for productions from Chinese and Japan to European and African market. Fig. 1 shows that the Singapore and the Malacca Straits are significant transshipment point and route for cargo originating from, or destined for, the European, East/Northeast Asian and Australasian markets in the world container flow. The Asia-Europe Route overtook the transpacific route as the largest containerized trading lane, with lane totaled 27.7 million TEUs in 2007.

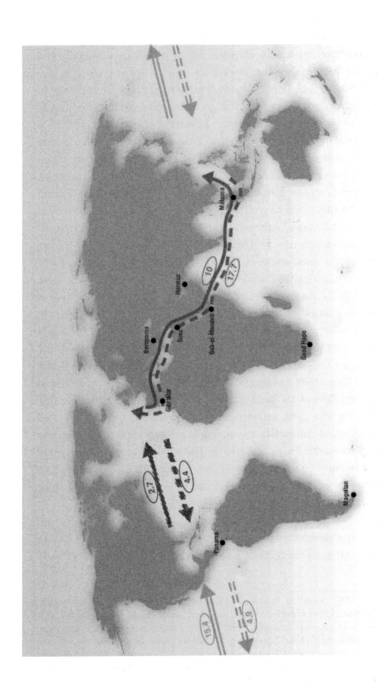

Fig. 1   Flow of container traffic in 2007 (Millions of TEUs), source from [UNCTAD, 2008].

According to [Klink and van den Berg, 1998] and [Heilling and Poister, 2000], ports are the most significant elements in maritime trade. The authors highlighted that role of ports as gateways to domestic and international trade, connecting the region as well as intra-region to the world is pivotal in global logistical network. Furthermore, the importance of ports to an economy cannot be underestimated, recognizing that the impact of having a competitive port is far reaching beyond the immediate benefits such as higher operating efficiency, profitability, competitive exports and employment opportunities. Being a vital link in the overall trading chain and consequently, port performances, to a large extent, determine a nation's international competitiveness. Prior to this, [Yabe, 1990] emphasized ports not only function as junctions of marine and land transportation but also as nucleus areas for industrial activities and cities[a] since ancient times.

Nonetheless, developments in the technological infrastructure have significantly altered the landscape of the port industry over these recent decades. Among these, logistical and communication advancements have led to hinterland expansions and overlaps and ports are no longer captive of their hinterlands. As carriers deploy larger vessels for higher cost efficiencies, the accompanying reduction of port calls made by these carriers increases the potent impact of a move of a carrier and fuels greater competition in the port industry. In order to insure that their ports will remain attractive in the heightened competition, port authorities will need to have a clear picture of the changing playing field of the port and maritime industry. These include an understanding on the port selection criteria adopted by carriers and the underlying factors of competitiveness. Ports are required to continually assess its performance relative to the rest of the world so that appropriate strategies can be devised to meet the ever increasing and more demanding needs of port users as well as maintain continuing competitiveness of their ports and economy [Tang et al., 2008].

In this chapter, we review previous research on maritime trade evolution in Asia and trace the development of major Asian ports, so as to validate previous findings and provide further insights on the development of port

---

[a][Fujita and Mori, 1996] observed that most East Asian countries (like Indonesia, Philippines and Thailand) had experienced a disproportionate share of population and manufacturing industries that are concentrated in their primal cities located at ports. The same goes for Shanghai, Hong Kong and Singapore. One key reason is because ports represent the most convenient location for exports and imports. For some large cities such as Chicago and Paris, even though ports do not play an important role today, their primal growth had been initiated by the ports.

cities. Specifically, the long tale of maritime trade evolution in Asia and development of Asian port cities seeks to unveil the relationship between maritime industry and the economic progress of a city. Recognizing that port forms a major pillar[b] in the development of a port city, we identify some of the key success factors and challenges faced by major ports in Asia. These identifications are subsequently verified with findings in the extant literature. It is hoped that such historical documentations supported by contemporary case studies will provide invaluable guidance for the future development in Asia port and maritime industry.

The remainder of this chapter is organized as follows: Section 1.2 reviews of maritime trade evolution in Asia from thirteenth century to post World War II. Section 1.3 documents the development of major Asian ports. Section 1.4 discusses some of the important factors that are found to influence the performances of ports in the classic literature. Section 1.5 highlights some insights and fruits for thoughts in terms of future development. Section 1.6 summarizes and concludes the chapter.

2. Evolution of Asia's Maritime Trade

Maritime trade and ports evolution are affected by revolutions in the transport sector and the industries, as well as, the globalization of the economy. In ancient times, international maritime is the major transport mode for geographical discoveries, and maritime trade plays an important role in civilizations of mankind. Following the industrial revolution, British traders extracted raw materials at low cost from their colonies, while dumping industry goods into these colonies markets. This was the key driving force behind maritime trade. After World War II, trade liberalization led to increased participations from developed and developing countries in international trade and fueled the growth of maritime activities. Subsequently, globalization in the 1990s had brought about a large expansion of world trade and shipping, of which, maritime trade has played an increasingly important role in stimulating economic growth.[c]

---

[b]Therefore, factors contributing to a successful port are important not only for port marketing strategy, but also for the strategies management for the spatial economic development of port cities.

[c][Irwin and Tervio, 2002] have proven one of the most fundamental propositions of international trade theory, which advocates that trade allows a country to achieve a higher real income than would otherwise be possible.

In the following sub-sections, we explore the evolution of Asia's maritime trade classified into three important distinct periods: ancient maritime trade between India and China prior to the fifteen century, maritime networks during times of colonization and Asia's maritime trade under globalization.

2.1. *Ancient maritime trade between India and China, 1200–1450*

Folk traders had established China and India maritime links from the first century BC. More notably, 1200–1450 marked a distinct milestone in the history of China and southern Asia maritime relations with the forming of the government maritime network. From the end of Song Dynasty to Ming Dynasty, Chinese government organized many fleets to southern India and even east Africa as parts of the Indian commercial zones. It was said in [Liu, 1988] that traders involved in these maritime exchanges are primarily the Persians, other Middle Easterners, South and Southeast Asians. The Malacca Straits, controlled by the port kingdom of Malacca at that time, was then already a critical trade route linking the Indian Ocean to the South China Sea and Pacific Ocean. Silk yarn entered India and was shipped to Rome through Indian ports, while coral and glass from Roman reached Chinese markets through Indian ports [Lin, 1998]. Hence, ports in southern Asia were important transit points for Chinese traders to Persian Gulf, and were also transition centers for Chinese and Roman goods. [Sen, 2006] noted that these maritime trades had grew so rapidly thereafter that the ports-of-trade in Southeast Asia and a Muslim trading network were formed in the eighth and ninth centuries respectively.

After Qubilai (the King of Yuan Dynasty) took control the ports in China, he executed an aggressive maritime policy with a desire to expand the military and political influences to the southern coastal region. Large Chinese ships,[d] which were more than thirty meters in length with capacity over hundred tons and staffed with at least sixty crews, were deployed to carry out the maritime trade in the early twelfth century. These ships sailed around Chinese sea and were capable of reaching the land of Korea and Japan by today's standards. [Sen, 2006] noted that significant fiscal revenue was derived through maritime commerce, which supported the

---

[d]Marco Polo, who visited China at Yuan Dynasty, described the ships that were transporting goods between China and India as ships having nailed hulls and multiple masts and cabins and were able to carry 1860 tons load.

world military expansion of Yuan Dynasty. Particularly, the maritime route through the southern Indian coasts was crucial for commerce between China and Persian Gulf. The associated significance of the south Indian ports to maritime trade during the thirteenth and fourteenth century is described in many history literatures, including [Grant, 2002].

Unlike Yuan Dynasty, the objective of the Ming court for developing maritime links was not to profit from the commercial exchanges between China and India coastal regions. Rather, the Ming court desired to use its naval power to spread its rhetoric civilizing, maintaining peace and order, and economic prosperity across regions. From 1405 to 1433, Ming court supported Zheng He's seven voyages to southern Asia, Red Sea, Indian coast and east Africa [Sen, 2006]. [Lin, 1998] asserted that the mission of Zheng He's first two voyages, in 1405 and 1407 respectively, was a solicitation of tributary. The Ming ruler, in turn, invited the envoys from Calicut and other foreign representatives to banquets, conferred titles and returned gifts. Meanwhile, the expeditions of Zheng He also increased the maritime exchanges between China and the kingdoms along the Indian coast sharply.

Owing to Ming court's prohibition to private overseas trades, many Chinese merchants resided at foreign ports. These Chinese communities in Southeast Asia had found it easier and more profitable to operate at Java, Malacca, or other ports in the region for their convenient access to India and South China. In addition, they benefited from participating in the tributary system of Ming court, and avoided trade competition with other foreign traders in the Indian ports. These Chinese traders travelled from the Southeast Asian ports with the northeastern winds between December and March, and returned with southwestern winds between April and August [Grant, 2002].

Notwithstanding the differing motivations, the influences of Yuan and Ming courts in the Indian Ocean world were far-reaching and were recognized by officials in the coastal regions of India. Based the analysis of [Sen, 2006], with City University of New York, the formation of maritime networks to Indian coast by private and official Chinese traders had augmented the domain knowledge on Indian geography, coastal kingdoms and commercial prospects under the ruling of both dynasties.

2.2. *Maritime networks in colony times, 1500–1950*

In 1511, Portugal captured Malacca for its strategic importance. In 1641, the Dutch occupied Jakarta, and established Dutch East India Company to control the trade in the Straits from the seventeenth to eighteenth century [Chua *et al.*, 2000]. After the industrial revolutions, the British also recognized that a safe passing of British cargos into the Chinese market could be ensured with a control over the Malacca Straits. Thus, in 1819, British established a colony in Singapore, and agreed to open the Straits for other friendly nations, which ended the long-standing dispute with Dutch. Singapore, Hong Kong and Calcutta ports had become colonial maritime centers serving a key global trading route. In the nineteenth century, these port cities were integrally linked by the East India Company in Asia. As the colonial port cities, Singapore, Hong Kong and Calcutta enjoyed rapid economic growth, physical transformation, ascending population numbers, and dense maritime networks connected by international trade [Tan, 2007]. Their ports played a crucial role not only as trading places, but also as centers for technological transfer and culture communication during the period of colonization. The following paragraphs shall focus on the discussions of each of these port cities in turn.

Singapore occupies a strategic location between the Indian Ocean and Pacific Ocean, and at the southernmost tip of Asian landmass. Owing to its geographical position, transshipments made up a large proportion of Singapore trade in the nineteenth century. Being part of a trading environment, the economic, social and cultural conditions of Singapore were determined by the flow of its maritime networks. During the colonial times, the extended commercial networks of Singapore were integrated by a full range of maritime vessels and formed by physical connections, maritime routes, functional inter-dependence such as trade, labor, commodity exchange and capital flows [Harper, 2002]. The colonial trading pattern promoted Singapore to be a critical node in the whole maritime network. It connected the Persian Gulf and India to the west with China to the east for centuries, which was named Maritime Silk Road. By the middle of nineteenth century, Singapore had become a congregation of multiple communities as Indians, Chinese, Malays and European trading together in the market.

In the late nineteenth century, the speed of globalization (generated by trade and imperialism) was accelerated by the advent of steam vessels and telegraph. During this time, ancient trans-national connections stretching from the Arab lands to the south Chinese coast were revitalized. Singapore became not only the key economic node but also the heart of intellectual world of Asia. The island housed a dynamic mixture of diverse classes and culture, triggering innovation with local experience and improving relationships with regional and international communities. With the opening of Malayan peninsula at the end of nineteenth century, Singapore acquired a physical hinterland from which the Malayan's agriculture productions were exported. These export activities had become a major driving force of the port traffic and Singapore economy by the end of nineteenth century.

Similar to Singapore, Hong Kong was a fishing coastal villages consisting of a hundred dwellers before the intervention of external powers. While advantageous locations and nautical accessibilities conferred both economies their strategic importance as British colonies, the main reason for becoming colonial city ports differed. According to [Lee *et al.*, 2008], it was Hong Kong's potential as a gateway to China that motivated the British Empire to establish and start trade negotiations in Hong Kong. However, the failure to dominate market in China had lead to the development of Hong Kong to inevitably parallel that of Singapore.

In contrast to Singapore and Hong Kong, Calcutta's early success as an international hub port city was owed to its ability to transform its immediate hinterland in northern India into an international market and not relying mainly on transshipment. This hinterland, which spanned from the Gangetic plains to the west and the Brahmaputra valley in the northeast of India, had provided Calcutta with large volumes of trade and labors. As all important offices moved from Murshidabad to Calcutta in 1772, Calcutta became the capital of British India [Tan, 2007]. Through the connection of the Indian hinterland to world market (especially China and Southeast Asia), Calcutta port delivered half of Indian's export of cotton, tea, coal, sugar and saltpeter in the late 18th century. For the whole 19th century, Calcutta was a centre of commerce, culture and administration. The transformation of Calcutta's hinterland into an international market was greatly aided by the opening of the Suez Canal and its special location on a navigable river that expanded its trade-related commerce, as well as, its extensive rail and road network that provided it with a large number of laborers and immigrants. Until the early 20th century, Calcutta was an international port and the center of colonial trading, serving the vast business created by the East Indian Company. However, from the 20th century, the British moved its capital of Indian Empire from Calcutta to New Delhi, and Calcutta gradually lost its hinterland market and its position as the empire port city.

2.3. *Asia's maritime trade under globalization*

After the World War II, a trading regime GATT (General Agreement on Trade Regime 1947) was established to govern trade between industrial countries. According to [Francois and Wooton, 2001], the Uruguay Round of World Trade Organization (WTO) in 1993 marked a new shift in the maritime trade system by involving a commitment from the developing countries to participate in the multilateral trading system. Thereafter, tariff barriers had further assuaged with the formation of more trade liberalization districts under the agreements of WTO. In addition, other shipping conferences are organized to set rates, analyze market conditions, assess other development such as fuel prices and port charges. One of these major conference agreements is the Transpacific Stabilization Agreement, which controlled about 86 percent of US maritime trade with Asia in 1998.

Between the 1950s and 1990s, the Asia economy saw the industrializations of large economic giants and the globalization of port cities. During the 1950s and 1960s, Japan embarked on an aggressive industrialization of its economy which was then followed by Korea some twenty years later. As part of these industrialization efforts, the Japanese and Korean governments gave strong support to lead industries that were tailored for exports. These include the shipbuilding, motor vehicles and electronics industries etc. Together with the huge amount of maritime imports of raw material and energy (such as metal ores and coal), the trade in Japan and Korea increased dramatically and stimulated traffic growths in the, Kobe, Osaka Tokyo, Yokohama and Busan ports. Particularly, [Yabe, 1991] noted that the emphasis on the roles of ports to the industries in Japan during the early industrialization period. Extension of wharves and large landfills for industrial areas were carried out in many ports and harbors, in response to the rapid increase in production and distribution. Many ports have been developed seaward so that sufficient water depths for larger ships and adequate areas for cargo handling can be created. Comparatively, urban life was neglected resulting in the occurrences of various problems such as water pollution, traffic congestion and loss of access to the waterfront for the people living in the city. In recognition of these, Japanese port development policy has been drastically changed to take into considerations of the effect on the city since 1985.

Subsequently, with the trend towards globalization, ports in Asia delineated free trade zones to boost their attractiveness and competitiveness as logistics hubs within world maritime trade networks. As ports continue to play an important role in economic development, many saw an expansion in capacity and upgrading of technology, aimed at improving port operations efficiency to attract more direct carrier services[e] and keep pace with the booming maritime trade caused by globalization. Through a series of advanced modernization and urban expansions, Singapore and Hong Kong had become two hub port cities, connecting Europe and North America with China and Southeast Asia. [Cullinane *et al.*, 2007] had proclaimed that Singapore posited itself as a global city-state, defining itself as a city with global orientations, and entrenched itself as hub for global international manufacturing, commerce, communications and finance networks. Together, the post-industrialization, globalization and China's Open Door Policy had provided new changes for Singapore and Hong Kong. Nevertheless, emphases on port productivity and efficiency improvements, urban attractiveness and a total port-city separation were not relaxed.

In summary, the Asia's maritime landscape had undergone three prominent phases of trade evolution (i.e., ancient official and private trade, colonial economic trade and globalization maritime trade) that produced profound impacts on the development of ports in Asia (see Fig. 2). Likewise,

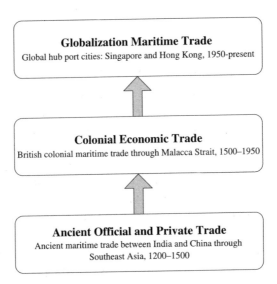

Fig. 2   Evolution of maritime trade in Asia.

---

[e]Direct carrier services augment the connectivity of a port by increasing the number of global destinations can be reached by shippers at the port and speeding transiting times.

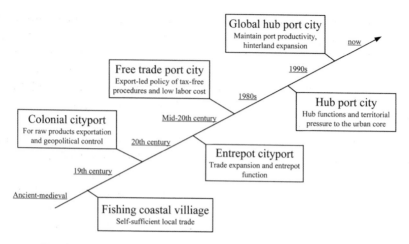

Fig. 3   Evolution of ports in Asia, modified from [Lee *et al.*, 2008].

the development of port city in Asia had followed a progression path from a fishing coastal village to colonial city port, to entrepot city port, to free trade port city and finally emerged as a hub port city, as illustrated in Fig. 3.

3. Asia Port Developments in the 1990s and Beyond

Since the 1990s, Asia has experienced rapid economic growth. Compared to the world gross domestic product (GDP) that is growing at an estimated rate of 3.1 percent in real terms, the aggregate economy of the East Asia maintains its upward momentum with a 7.0 percent growth rate in 2008. Despite the contraction of the Japanese economy,[f] China and India have shown remarkable growth of 9.6 percent and 7.4 percent respectively while South Korea grew by 2.2 percent in 2008.

During the same period, the world container port throughput grows by 4.06 percent to over 507 million TEUs in 2008 (UNCTAD, 2009). Meanwhile, some ports in Asia have reported double-digit gains. These ports include Ningbo (19.94 percent), Tianjin (19.67 percent), Guangzhou (19.58 percent), Port Klang (11.96 percent), and Dubai (11.02 percent). On the average, the mainland Chinese ports grew by 9.42 percent.

---

[f] Japan has experienced a negative GDP growth of −0.7% in 2008.

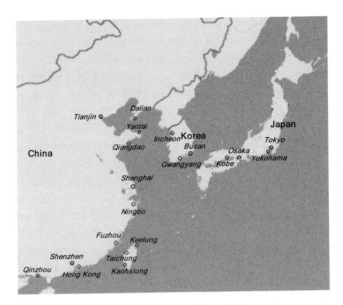

Fig. 4   Ports in Northeast Asia.

3.1. *Northeast Asia*

The following subsection describes the key advantages facilitating the growth of some of the major container ports in China (including Hong Kong and Taiwan), South Korea and Japan, as well as, development plans to counteract the challenges facing these ports in their future growths. Figure 4 illustrates the major ports in Northeast Asia.

3.1.1. *Greater China region*

The Hong Kong port is the world's 3rd busiest ports. Located on the north shore of the South China Sea at the mouth of the Pearl River Delta, the port of Hong Kong is the leading container port for the mainland of China and a major hub port for intra-Asia trade. [Wang, 1998] advocated that Hong Kong's proximity to underdeveloped Chinese ports is one of the prime reasons that had allowed Hong Kong to achieve its load center status in a very short period of time. [Cullinane *et al.*, 2004] added that highly educated workforce in Hong Kong is another factor that promoted the Hong Kong Port's international status as a major hub port in Asia. Being at the centre of the Asia — Pacific Basin and strategically placed on the Far East trade routes, the port has also been a key factor in the development of the area.

The Hong Kong port is well-regarded as a highly efficient international container port in the world. It possesses one of the most perfect natural harbours in the world and operates in a business-friendly environment with world-class infrastructure, The port, handling 24,248,000 twenty-foot equivalent units (TEUs) of containers in 2008, is served by 192,000 vessels from some 80 international shipping lines that provide over 450 container liner services per week connecting to over 500 destinations worldwide.

However, Wang indicated that Hong Kong port faced two dimensions regarding the space problems: first, the lack of stacking space within the port; second, the lack of stacking, parking, and repairing space outside the port. Furthermore, [Loo, 2002] observed that the abundance of labor in China has led to a large-scale relocation of labor-intensive and export-oriented industries into China, which spurred the growth of ports in South China. As operations in the Chinese ports improve, the differential advantage in terms of efficiency at Hong Kong port will be gradually eroded. While Hong Kong is still the leader in terms of value-added trade services such as consolidation, forwarding and financing, the cost advantage of its adjacent Shenzhen port and other ports in Southern China represents a constant threat. Statistics reveal that the Hong Kong port has lost as much as 40 percent of its monopolized traffic from the region in the 1990's to ports in Southern China. In response to these challenges, Hong Kong has taken some measures to further enhance port productivity and efficiency as well as setting up high technical logistics centers and open space (OS) zones.

Spacious water areas, developed hinterland and convenient land transport links are some of the advantages that had contributed to the early development of Kaohsiung port. Situated in the South-Western part of Taiwan at the nexus of main Asia Pacific trade routes, the naturally deep-water port derives 52.2 percent of its volume from transshipment and enjoys low tidal variance. The port also has ample space for expansion and provides one of the world's largest ship scrapping facilities. However, [Haynes *et al.*, 1997] noted that the total cargo and containerized cargo growth in Kaohsiung (Taiwan's largest port) has been lagging behind Hong Kong and Singapore due to customers' dissatisfactions with service such as cumbersome custom clearances, high costs and poor management. In 2008, the port of Kaohsiung received 36,000 vessel calls and handled 9,677,000 TEUs which puts the port in the 9th and 12th position respectively in the ranking of Asia's and world's ports. The most recent statistics reveal that the port served 8,102 incoming container vessels, of which, 1,584 are over

60,000 tons in 2009. In line with the Taiwanese government's strategy to promote Kaohsiung into an Asia Pacific operation headquarter, a privatized Free Trade Zone has been established and operated since 2006. The Free Trade Zone has vast area of adjoining land, which serves as an offshore shipping center or logistics center that supports a combination of repacking; processing and other value-added functions.

The port of Keelung, another major Taiwanese port, lies on the northern part of Taiwan, 40 km away from the capital city of Taipei. A series of dredging programs have been undertaken to serve the increasing number of larger vessels and attract liners to call upon Keelung Port. After the completion of the first two phases of the intended dredging programs in 2001, the approximate depth of main channel and the diameter of turning basin are 15.5 meters and 650 meters, respectively. Currently, these dimensions are sufficient to meet the berthing needs of 60,000 tonnage container vessels. In addition, the Keelung port also establishes affiliations with the ports of Oakland, Los Angeles, Bellingham and San Francisco in the United States and the Port of Southampton in the United Kingdom as "sister ports" for purpose of promoting international friendship and strengthening the exchange of technology and experiences on port developments. These port development efforts are seen to have paid off from the recorded container traffic of 2,128,000 TEUs in 2008. With an average annual port call of 9,200 vessels, the port has shipping routes linking globally with all the other major container ports.

Among the main container ports of Taiwan, Taichung is the closest to mainland China. The port of Taichung, which is called upon by 5,950 vessels, is located on the west coast of Taiwan with a total length of 12.5 km and widths between 2.5 and 4.5 km. While container traffic is on a much smaller scale (i.e., approximately 1,200,000 TEUs), it has grown the most rapidly among all the Taiwanese ports. Owing to the fast increase in business volume, a construction plan to expand the port was carried out in cooperation with Taiwan Construction. Given the good investment environment, fully automated warehouse operations and high service efficiency, the port of Taichung is potentially the main contender for direct trading links between Taiwan and mainland China.

China ports can be divided into Northeast China ports, east China ports and south China ports. As a new active economic driver in Asia and the world's largest container generator since its trade liberalizations, China has dramatically increase its financial investment to improve the infrastructure and superstructure in many of the Chinese container ports.

In 2007, Chinese ports alone accounted for 139.1 million TEU, which represent 28.4% of world container port throughput [UNCTAD, 2008].

Because nearly 50 percent of all foreign investments into China are devoted into Shanghai and its surrounding, the port of Shanghai alone takes nearly half of the total container traffic through all ports in China making it the leading Chinese container port. The port of Shanghai is located at the mouth of Changjiang (Yangtze) on the apex of a vast hinterland with inter-modal waterways, rail and road links running inland to central China. Containerization is high at Shanghai port, reaching above 55 percent, and the port is attracting an increasing number of direct, deep-sea vessel calls. The 2008 statistics show that a total of 27,980,000 TEUs from 55,000 vessels passed through the port of Shanghai, and this impressive volume has placed Shanghai as the second busiest port in the world. Alongside, Shanghai and its surrounding provinces have become the driving engine of China's economy growth. Particularly, Shanghai functions as the economic, financial and shipping centre of China, with its surrounding provinces as the centre of manufacturing.

To some extent, the rapid ascend of the Shanghai port on the world rankings can be attributed to a series of aggressive port development efforts (particularly, with regards to the sustained investment in new terminals) embarked by its port authority. For instances, the second phase of the modernization of Yangshan Port (offshore of Shanghai port) has seen an installation of 13 double-decker conveyors[g] that can handle two 20 or 40 feet containers at the same time. The completion of the third Phase that involves the development of a deep water port at Yangshan in 2009, helps to partly overcome the problem of shallow draught in Shanghai port that has previously limited the size of the vessels calling at the Shanghai port and volume of cargo they ship. Currently, the bulk of Shanghai's cargo originates in or travels to the conurbation and neighboring provinces of Jiangsu and Zhejiang. A coastal and inland container hub is being developed at Longwugang in Shanghai Harbor to extend the port's hinterland. On the softer aspects, the port authority has introduced world-class port management practices into the port of Shanghai. These include but not limited to the simplifications of custom procedures, implementation of computer linkage between the port, customs and other related agencies etc.

---

[g]Since the double-decker conveyors are implemented in 2008, the port's load or unload efficiency is 850.53 TEU/hour, and the conveying speed of a single conveyor is 123.16 TEU/hour.

The next major port in China is the Shenzhen port, which handles 21,414,000 TEUs in 2008. Ranking as the fourth busiest port in the world, Shenzhen port is an agglomeration of several ports including Yantian, Shekou, Chiwan and other smaller ports in Southern China's Guangdong province. Of these, Yantian is biggest and sited in the sheltered waters on Dapeng Bay just 20 nautical miles north of Hong Kong. The port of Yantian is opened in July 1994 as an alternative access point to Southern China. Since its opening, this deep-water container port has lifted its throughput to over 1 million TEU in just 4 years and over 8 million by 2008. The main bulk of Yantian's cargoes originated largely from Shenzhen, Dongguan, Guangzhou, Huizhou and other Pearl River locations. The port is equipped with advanced port facilities and is served by a sophisticated rail and road network. Meanwhile, Chiwan and Shekou international container terminals have constantly improved the efficiency of custom procedures. More international shipping companies are choosing the Shenzhen port for transshipment due to its merits such as lower costs[h] and simplified customs procedures. By offering shorter time and lower cost of transport (including handling charges) between Hong Kong and the rest of China, the port of Shenzhen port has now become the 2nd largest port on the Chinese mainland in terms of handling international transshipment goods.

Other Chinese ports with impressive traffic performances are Ningbo (11,226,000 TEUs), Guangzhou (11,001,000), Qingdao (10,320,000) and Tianjin (8,500,000), which ranked 7th, 8th, 10th and 14th respectively in the world. Another newly developed port is Qinzhou in Guangxi province. Qinzhou port is the 6th bonded port in China. The port has 22 berths and a capacity of 15 million tons. Among the 22 berths, 11 can handle ships above 10,000 tons. Currently, another 9 berths providing an additional 25 million tons, are under construction. With all these recent developments, earlier observations from [Cullinane *et al.*, 2004] and others that Chinese ports were under-provisions of physical infrastructure resulting long waiting time is no longer valid. With logistics infrastructure and management knowhow, the future of the Chinese ports is optimistic.

3.1.2. *Korea*

The Busan port is by far the most important container port in South Korea, accounting for more than 90 percent of the nation's container throughput.

---

[h]The Shenzhen port's loading and unloading charges are low, nearly half of those at ports in neighboring Hong Kong.

Located at the eastern tip of the Korean Peninsula, the gateway on transpacific route, Busan port includes four branches: North Harbour, South Harbour, Gamcheon Harbour and Dadepo Harbour. As of 2008, Busan port is the 5th busiest port in the world with a container throughput of 13,425,000 TEUs and 83,547 vessel calls. Transshipment cargo accounts for some 43.2 percent of container throughput, while inbound and outbound cargo accounts for 28.5 and 28.3 percent respectively. A number of factors have contributed to the growth of Busan port. Firstly, Busan is a natural deep water harbor which allows the berthing of big vessels. Secondly, Busan is at the cross road of Northeast China, Japan and Western Russia and thus has potential to be the regional hub. Third, Busan is an attractive relay centre for minor Japanese ports because it is cheaper and thus able to undercut major Japanese ports. Thus, many shippers have been sending their cargos through Busan for transshipment to/from regional Japanese ports. Busan is planning build a total of 30 new berths for 50,000 ton ships by 2011 as competitions from Chinese ports (such as Shanghai and Dalian) intensify. With the addition of the new berths, the annual handling capacity of Busan is expected to reach 8 million TEUs.

Other than Busan, the Port of Incheon has contributed greatly to the development of the economy and industries in South Korea. Located on the mid-western coast of the Korean Peninsula, Port of Incheon is a gateway to Seoul. As an artificial port with the world's largest and most advanced lock gate (wet dock) facilities that overcome a tidal difference of 10 meters and permit vessels up to 50,000 DWT to berth directly in the inner closed harbor basin, the port is also equipped with various modernized harbor facilities for trade promotion with the main ports of the world. Nonetheless, the container traffic at the Port of Incheon is merely over 10 percent of that in Busan (i.e., 1,655,500 TEUs).

Going down south, Port of Gwangyang is situated on the south coast of South Korea above the Gwangyang Ha River of Yosu. The port of Gwangyang is equipped with an annual capacity of 5 million TEU and is the fastest expanding port in Korea. Since 1998, the port has been operating three branches — West, East and Yulchon Harbour. The port is connected to land through four eastern and western container driveways, and directly to a 2.5 km railroad. Yeosu Airport, which is near the Port of Gwangyang, is currently under expansion. Thus, a systematical network that enables fast commuting in every direction to and fro the port is formed with the integration of railroad, highway and other private airports. In terms of future developments, the port of Gwangyang is scheduled to be developed into a 33-berth super-scale container port by 2011 that is capable of handling 9.3 million TEUs annually.

However, [Bong, 2009] warned that growth rate of cargo traffic through Korean ports is drastically slowing down from 13.8 percent in the 1990s to 2.2 percent in 2008. Bong advocated that several causes of slowdown in cargo traffic through ports in Korea are: (i) a rise in operating cost, (ii) a slowdown in foreign trade due to change of industrial structure (i.e., increased portion of service industry), (iii) loss of competitive advantage in traditional labor intensive light industries, and (iv) failure to induce newly emerging high technology capital intensive industries. The author also suggested that Korea should capitalize on the underdeveloped logistics sector, so as to create more value-add through implementations of multi-modal transportation systems, technology and management know-how that drives the growth of ports.

3.1.3. *Japan*

Kobe, Osaka, Tokyo and Yokohama ports represent four of the major ports in Japan. Specifically, Kobe Osaka, Tokyo and Yokohama ports have experienced container traffic of 2,432,000 TEUs, 1,725,500 TEUs, 4,271,000 TEUs and 3,490,000 TEUs, respectively in 2008.

Kobe port is located in the central part of the Japanese Archipelago. Originating as a hub of trading between Japan and the Chinese continent and Korean peninsula during ancient and medieval times, Kobe port has a hinterland that covers the whole of western Japan. The geography location and topology have conferred Kobe port several unique advantages that makes it the principal foreign trade port of Japan. Firstly, Kobe port lies on the main routes of world marine-transportation networks. Secondly, Kobe port is accessible from various directions as it stretches from east to west. Thirdly, expensive dredging is unnecessary owing to favorable natural conditions that include some deep-waters berth and no seasonal winds and rivers flow into the port. Fourthly, the port is also ideal for mooring since it has little variation in tides. In terms of connectivity, the Kobe port is served by many regular carrier service lines, including North American, European, Southeast Asian, and Chinese lines that linked the port with 500 ports in 130 countries. 2006 marks the opening of Kobe Airport with the provision of Kobe-Kanku Bay Shuttle that provides a ferry service between Kobe Port and Kansai International Airport. Together with the existing expressway networks, domestic feeder services, and ferry services, intermodal transportation efficiency is secured. As part of a continual improvement process, Kobe seeks to constantly enhance its services for user convenience and friendliness by reducing port facility charges, simplifying various port procedures, computerizing operations using EDI (electronic data interchange) system for submitting various application. In order to provide greater flexibility to carriers, domestic container feeders are also permitted to use overseas berths.

The port of Osaka is located in the western part of the city of Osaka. Similar to Kobe port, the port of Osaka is directly connected to the main area of the country through an advanced network of expressways and other main roads as well as a feeder network. It is also directly linked up with Kansai International airport. The port possesses wharf facilities, 11 berths and over 3000 meters of quay with draft between 10 and 14 meters in the Sakishima District. Meanwhile, the terminals in the Yumeshima District offer 3 berths and over 1100 meters of quay 15 meters with draft of 15 meters. Port of Osaka is called by more than 7,000 ocean-going vessels per year, of which, more than 5,000 are container carriers. Over the years, the port enjoys increasing volume of container cargo, particularly with Asia, as a result of its constant attempt to promote user-friendly services to the users.

The Port of Tokyo is located on the west coast of Honshu in area between the estuaries of the Arakawa and Tamagawa Rivers. In 2007, the port served 31,332 incoming vessels and handled 87.63 million ton of cargos. The port plays an important role in the distribution of essential commodities such as sundry goods, foodstuffs, paper products, building materials and so forth throughout the Tokyo Metropolitan area (Shinetsu and southern Tohoku) for its industrial activities and 40 million citizens. Hence, the port has taken early actions to enhance the accessibility and functionality of its terminals for container, ferry and specialized cargo use. For examples, warehouses and distribution centers have been set up in the reclamation areas behind each terminal to complement terminal functions. Arterial routes, rail and other roadways are developed to facilitate distribution activities. There are also plans to construct new container terminals in the Outer Central Breakwater Reclamation Area to serve the key routes from Asia.

Close to the port of Tokyo, the Port of Yokohama is located on the northwestern edge of Tokyo Bay, 30 km from Tokyo. The Port of Yokohama is a naturally blessed port with a spacious water area of ample depth on the eastern side and undulated hills on the northern, western and southern sides. The port operates 24 hours daily and has been equipped with various facilities such as inner and outer breakwaters to protect the port from the effects of winds and tides. As part of the future plans, Japanese government aspires to develop the Port of Yokohama into a major container hub port, with separate facilities for intercontinental and Asian container traffic. It is estimated that the total cargo volume in 2015 will reach 150 million tons and the number of cargos handled will exceed 4 million TEUs.

Operating in a country that is technologically and economically advanced, all Japanese ports are able to offer efficient services through extensive use of sophisticated state-of-the-art facilities and implementation of modern management practices. However, apart from the generally more expensive labor and associated operations cost, [Imai *et al.*, 2001] pointed out that charges in Japan's ports have been consistently higher than those in other major hubs owing to overcapitalization of the port for relatively small cargo volume.

3.2. *Southeast Asia*

The following subsection describes the key advantages and challenges facing some of the major ports in Singapore, Malaysia, Indonesia, Philippines and Thailand. Figure 5 illustrates major ports in Southeast Asia.

3.2.1. *Singapore*

The Singapore port is strategically positioned to participate as a transhipment hub for South East Asia and contribute significantly to the country's growth[i] process into one of the core global cities in Asia. Specifically, the Singapore port is located at the crossroads of international trading in sea routes in the Asia-Pacific where the geographical topology endows the port with a naturally deep harbour. The port represents an active feeder shipping spot in Asia, with a network service ranging from short to long routes. Other than being highly efficient, the port offer full range of service, including fuel, pilotage and towage, cargo, vessel repairs, warehousing, banking, insurance, communications, entertainment, training and education in port operation and management, logistics and distribution management and other transport studies.

---

[i]The maritime industry comprises more that 5,000 establishments, employs around 100,000 people and contributes more than 7 percent of the Singapore's GDP in 2008.

24

X. J. Yang et al.

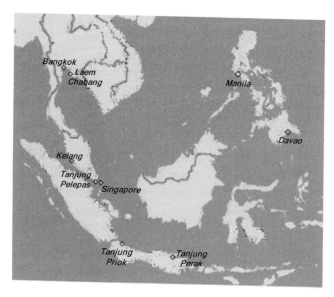

Fig. 5   Ports in Southeast Asia.

Throughout many decades, the Singapore port has retained her position as one of the world's busiest ports in terms of vessel arrivals, bunker sales, cargo tonnage handled and container throughput [Cullinane et al., 2006]. The 'secrets' of Singapore port's success are well-documented in commentaries and academic studies. Among these, [Zhu et al., 2002] argued that conductive Singapore's business environments and well-developed infrastructures are the main factors attracting MNCs investments. The traffic at the Singapore port is further augmented through the port-related industries, which are located in dense and compact districts and high technical logistic centers as a response to global and local forces that promotes in and outward multi-national operations. Other reasons for Singapore port's success can be attributed primarily to the resident port and maritime-related community which provide competitive products and top service standards in world-class to meet the requirements of port customers.

Today, Singapore port has achieved an impressive container throughput of 27,900,000 TEUs and become a focal point for 174,620 vessels of some 200 shipping lines with links to more than 600 ports in over 120 countries worldwide. Singapore port has 49 berths serving container ships, which can handle up to 26.1 million twenty-foot-equivalent units (TEUs). The construction of 16 berths has begun in October 2007, when completed in

2013, the port will have an annual handling capacity of 14 million standard containers (which is an increase of more than 50 percent). In large part, Singapore's historical importance was due to its geographic position in relation to the Straits of Malacca, one of the world's busiest sea-lanes. While the port of Singapore continues to serve as an important link for goods shipped between Asia and Europe, the port has been faced with stiff competition as an international transportation hub from neighboring Malaysia in these recent years. Malaysia started taking away Singapore's container trade business with the opening of its Port of Tanjung Pelepas and immediately secured two of Singapore's biggest shipping clients. [Lam and Yap, 2006] conducted a comparison on the cost competitiveness among the port of Singapore, Port Kelang and Tanjung Pelapse port.

3.2.2. *Malaysia*

Port Kelang and Port of Tanjung Pelepas are two major ports in Malaysia. Port Kelang is the one of the most established ports in Malaysia, with a container traffic of 7,970,000 TEUs that ranks 15th in the world and 11th in Asia. Situated on the west coast of Peninsular Malaysia (40 km from the capital Kuala Lumpur), Port Klang's proximity to the greater Kelang Valley[j] makes it a premier port in Malaysia. The port has trade connections with over 120 countries and dealings with more than 500 ports around the world. It serves as the nation's load centre and regional transshipment centre, and is called upon by 17,000 vessels annually. Port efficiency is ensured through modern infrastructure facilities, computer information systems (including EDI), pre-clearance and advanced pre-clearance on Customs, Health and Immigration formalities.

The major thrust of Port Kelang's developments will be more industrial-based dealing with very large consignments, which are in line with the economic growth in the central region of the country and its identity as a regional transshipment base. Currently, the port authority is constructing additional facilities as part of its supply-driven policies. When Westport is completed, the facilities at Port Kelang will be sufficient to handle the projected cargo throughput 130.5 million tonnes (i.e., 8.4 million TEUs) at the end of 2010. As part of the future development plan, Port Kelang

---

[j]The Kelang Valley is the commercial and industrial hub of Malaysia as well as the country's most populous region.

will see further expansion of port facilities south of Port Kelang between Tanjung Rhu and Batu Laut (30 km from Port Kelang).

Port of Tanjung Pelepas (PTP) starts operations in October 1999 and aspires to be the region's premier transshipment hub. The port is located at the confluence of major shipping routes at the southern tip of Johor West in Malaysia. Being only 45 minutes from the confluence of the world's busiest shipping lanes, the PTP has steadily attracted the world's leading main shipping lines which include Maersk Sealand in 2000 and Evergreen Marine Corporation in 2002. Port traffic statistics shows that 3,368 vessels had stopped at the PTP and brought 5,600,000 TEUs of container traffic[k] to the port in 2008. This put the port into the 18th place in the world; and 12th place in Asia just behind Port Kelang. Factors that have contributed to rapid growth in the PTP are its excellent port facilities and infrastructure, supported by a state of the art integrated information technology systems and highly trained staff, which enabled high efficiency and productivity to be achieved. The 15 meters naturally sheltered deep-water port also boosts of its excellent connectivity via road, rail, air or sea. PTP currently has 12 berths and a terminal-handling capacity of 10 million TEUs. Under the existing expansion plan, the port would build eight new berths and include land reclamation and dredging. The long-term plan is to have 95 berths such that capacity will reach 150 million TEUs.

3.2.3. *Indonesia*

Indonesia has two principal ports, namely the Tanjung Priok and Tanjung Perak ports. Tanjung Priok port (also known as Jakarta's port) is located in western Java 13 km from the city centre of Jakarta. Tanjung Priok port is the main port for the major manufacturing region around Jakarta and west Java, and deals with both coastal and international trade. The port is constructed after the independence of the Indonesia Republic with the main purpose of ships' loading/unloading among the islands on recognition that the existing Sunda Kelapa Port was unable to be further developed to accommodate increasing trade ships brought about by the opening of Suez Canal. The Tanjung Priok port is well protected by breakwaters, with

---

[k]This figure translates into 8 percent of South-East Asia's total port market. Of the 5,600,000 TEUs, 95 percent are transshipment and 5 percent hinterland (i.e., local cargo). The port hopes to increase the latter to 20 percent in the short- to medium-term.

facilities for all types of cargoes. However, the port owner[1] noted that the growth of the port has been hampered by limited capacity and inefficient operations, including poor road access. Its current capacity is only 600 containers per hour, about two-thirds of the volume projected for next year.

Currently, the Tanjung Priok port ranks 26th in the world and 18th in Asia by handling 3,984,000 TEUs (which is about 30 percent of total freight handled by Tanjung Priok Port). In terms of vessel traffic, the port registered a total of 7,150 calls in 2008. The government has set an ambitious goal of developing an international-standard regional port. In order to fulfill this mission, the state-owned port operator has announced its plans to invest US$286.2 million to improve infrastructure (i.e., the purchase of new equipment and stronger cranes, and redesigning the docks) at Tanjung Priok port in Feb 2010. The investment is part of a five-year plan to modernize Tanjung Priok, cut costs in half and reduce the ship docking-time from three or four days to two days.

Meanwhile, the Port of Tanjung Perak (also known as the port of Surabaya City) is located on the northern coast of the island of Eastern Java, opposite Madura. The port serves 4,700 vessels as one of the main gateway ports to Indonesia. Being the principal port in East Java with an annual container traffic of 1,000,000 TEUs, the port also functions as a main cargo collection and distribution center for both the Province of East Java, and the whole eastern archipelago of Indonesia. The Port of Tanjung Perak, equipped to accommodate tankers, general cargo vessels and container vessels, has undergone continual physical development with modification of existing berths, and provision of additional berths specifically designed for container handling operations. The Port Authority, in its efforts to encourage development of the associated port industries and construction of the passenger terminal, continues to upgrade and improve both port facilities and services to meet demand.

3.2.4. *Philippines*

The two primary container ports in Philippines are Manila and Davao. Manila port, situated at the East end of Manila Bay, is the most significant port in Philippines. The port handles 4,062,000 TEUs, which accounts for over 90 percent of the nation's international cargoes. With regards to its

---

[1]The port operator JICT is jointly owned by Hong Kong's Hutchison Port Holdings and state-owned port operator PT Pelabuhan Indonesia II.

global and regional standings, Manila port is the 25th busiest in the world and 17th seventeenth in Asia. The port of Manila has significantly benefited from its topology that bestows it a shoreline of 2 km and protected by 3050m of rock barriers, while enclosing approximately 600 hectares of anchorage. The port of Manila is presently equipped with an annual capacity of 1.5 millions TEUs. Press release on April 2009 said that the Philippine government is forging ahead with plans to establish some 70 modular ro-ro ports over a period of four years to the tune of US$248.13 million. The proposed facilities call for the use of prefabricated steel modules for the port super-structure — pier, causeway, mooring platform, ramp dolphin and terminal with solar-powered utilities. Negotiations are under way for the requisite loan package, estimated to cover nearly 88 percent of total project cost.

The Davao port, situated on the southeastern coast of Mindanao Island, is the second largest port in Philippines with container traffic of 72,000 TEUs. The Port of Davao, otherwise known as Sasa Wharf, holds the distinction as the premier export and import hub in Mindanao. In addition, Davao port also functions as the front-line port in the exchange of commerce and trade between provinces and other parts of the country, as well as, the principal seaport for most commodities produced along the Davao Gulf. In terms of topological advantage, the natural islands of Samal and Talikod along Pakiputan Strait of Davao Gulf bound the port in the east. Hence, Davao port is relatively protected by landmasses on all sides except at the South. In December 2008, Davao port has completed a rehabilitation project involving the construction of a new 42.35 m × 18 m quay, an expansion of the 3,179 square meter Reinforced Concrete Wharf, 13,180 square meter back-up area, mooring and fendering area, drainage system, and the installations of port lighting and rockworks. The port can now accommodate eight ships at the same time.

3.2.5. *Thailand*

Bangkok and Laem Chabang are two of the leading ports in Thailand. Bangkok Port (also known as Krung Thep and Klong Toey) is located on the left side of the Chao Phraya River between 26.5 and 28.5 km from Klongtoey District, Bangkok. The port serves 2,800 vessels and handles 1,480,000 TEUs in 2008. Bangkok Port is well-connected with road and rail systems, which enable fast and economical transport of cargoes between the port and its hinterland. It is also equipped with a bonded warehouse that offers several value-added services such as online inventory account reporting, more equipment for lifting and moving goods, and expansion of storage areas. To provide a more comprehensive support for port-related activities and ensure optimal resource utilization, four zones are set out for future development. Specifically, Zone 1 will be developed into a Maritime Business Centre to accommodate shipping agents, freight forwarders, a Maritime Training Centre, an Educational and Exhibition Center, a Business Center catering to banks and financial institutions, and an Integrated IT/EDI Document Center; Zone 2 is set aside for the construction of high-rise logistics center, cargo consolidation/distribution center, tax free zone, general/bonded warehouses, and a Truck Terminal; Zone 3 will serve as a modern market that complies with sanitation and environmental standards; and Zone 4 will house a modern office building and other relevant activities.

Laem Chabang port is located on eastern Thailand in the Sriracha district, about 130 km south of Bangkok and Thailand's industrial heartland. The deep water harbors of the Laem Chabang port is opened in 1999. The port provides a comprehensive range of 24-7 services to exporters and importers. Further improvement to transport links has increased the accessibility of the port and the port has witnessed steadily rising traffic volume since its opening. Today, approximately 5,134,000 TEUs on 4,650 vessels go through Laem Chabang port making it the 21st busiest in the world and 15th in Asia. The current plan is to develop Laem Chabang into an e-Port. With a primary objective of relieving traffic bottleneck, the main features of this e-port are: (i) RFID-enabled payment system to inspect vehicles passing through e-Gate Control, verify data on the number of containers and fee collection; and (ii) Electronic Data Center for Real Time exchange of e-manifest, Container List and other data between Laem Chabang Port, dock operators, Customs and Immigration Department. Also, in Feb 2010, the National news bureau of Thailand publically announced that The Port Authority of Thailand will push forward Laem Chabang Port to become the main port for international and Mekong Sub Region trade, as well as, the center of trade links, business traffic, merchant marine. A four years development plan has been drawn up to increase the capacity and energy efficiency of the port so as to accommodate the rapid increase of the economy, trade and the growth of Thailand's international seaborne trade. When completed in 2016, Laem Chabang port will provide a capacity of up to 7.2 million TEUs a year.

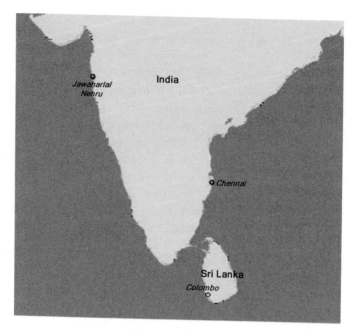

Fig. 6   Ports in South Asia.

3.3. *South Asia ports*

Compared to ports in Northeast and Southeast Asia, ports in South Asia are generally less developed and handle less container traffic. Exceptions are the major Indian and Sri Lankan ports. The following subsection gives brief accounts of the Jawaharlal Nehru, Chennai and Colombo ports. Figure 6 illustrates major ports in South Asia.

3.3.1. *India ports*

Jawaharlal Nehru and Chennai are major Indian ports. Jawaharlal Nehru port, also known as Mumbai port, is the biggest and most environmental friendly port in India. Commissioned in 1989, the port handles 3,953,000 TEUs (which translates into 55 to 60 percent of the nation's total containerized cargo) in 2008. This volume of container throughput places Jawaharlal Nehru port on 27th and 19th positions in worldwide and Asia. Jawaharlal Nehru port is well connected to the national extensive network of railways. The port operates 24 hours per day, possesses modern handling facilities and adopts up-to-date customs EDI and vessel traffic management system. In order to cope with the expected growth in container traffic and container vessel sizes, projects to expand ports facilities and further improve its rail and road connectivity are underway. More specially, these include: (i) the development of a standalone container handling facility with a quay length of 330 m towards North at JNPT; (ii) development of Fourth Terminal and Marine Chemical Terminal; (iii) deepening and widening of main harbor channel and JN Port channel; (iv) doubling of rail track; (v) development of an integrated and centralized tractor parking zone; and (vi) upgrading of existing roads and yards. On and all, when completed in 2011, the expanded port will assume an additional capacity of 0.8 million TEUs per annum from the optimum utilizations of the water front area and accommodate up to 14 m draught, while providing a safer movement for cargo and avoiding difficulties in container stacking which together enable faster turnaround time.

At the Coromandel Coast in South-East India, the Chennai Port (previously known as Madras) operates on a much smaller scale with container throughput totaling to 1,128,000 TEUs. Since 2009, the port operator has started the construction of a Second Container Terminal which features a 832 m long berth with 15.5 meters alongside depth and a back up area 35 hectares. Capacity of Terminal is 1 million TEUs per annum. Other efforts to modernize Chennai Port are the realignment of rail and road network inside the harbor in progress, mechanization of the coal conveyor system, and the deepening of channels, basins and berths. In addition, Chennai Port is contemplating to carry out the construction of a dedicated Elevated Expressway from Chennai Port to Maduravoyal, runs for a length of 19.01 kilometers along the riverbank and followed by the NH 4 road. This project is likely to be commissioned in February 2012. Through these series of continuous modernization that enable the provision of cost-effective and efficient services, as well as, the implementation of simple and integrated procedures, and user-friendly approach, the port seeks to achieve greater heights in the near future.

However, the growths of the Indian ports have generally lagged behind other ports in Asia. [De Monie, 1995] and [Haralambides and Behrens, 2000] cited poor physical configurations of ports, proximity to urban development, outdated port facilities, insufficient equipment maintenance, backward cargo handling techniques, inadequate accountability for cargo handling, bureaucratic administration, regulations and weak coordination between departments as possible reasons for poor performance of Indian ports. [De and Ghosh, 2003] confirmed the hypothesis that if an Indian port can perform better by improving its operational and asset performances, then it is likely to get higher traffic but the reverse is untrue.

3.3.2. Sri Lanka ports

The port of Colombo lies on Sri Lanka's southwestern shores on the Kelani River. Apart from being a major port in the Indian Ocean, the Port of Colombo also makes it presence felt as the world's 28th and Asia's 20th busiest container ports with a registered traffic of 3,687,000 TEUs. The port handles most of the Sri Lankan's foreign trade, including the manufacturing exports of processed raw materials. Other leading industries in the Port of Colombo include jewelry, chemicals, glass, textiles, leather goods, cement, and furniture. As the commercial center of Sri Lanka, the Port of Colombo also contains head offices of both foreign and local banks, insurance companies, government offices, and brokerage houses.

With one of the world's biggest artificial harbors, the Port of Colombo is made up of three terminals and offers a total of 6245 square meters of bonded warehouse including 125 square meters of cool room. These warehouses are equipped to accept all types of goods (except dangerous or perishable goods), with 24 hr security service provided, and small processing services for re-export cargoes available at Warehouse BQII. The Port of Colombo is presently undergoing an expansion on west of its current southwest breakwater. According to the Sri Lanka Ports authority, the South Harbor Development Project involves four terminals. Each of this four terminals will be over 1200 meters long with alongside depths of 18 meters and covers about 600 hectares in total. There are future plans to deepen the berths in the Port of Colombo to 23 meters for deep-draft vessels. The South Harbor channel will be 560 meters long with a depth of 20 meters and a harbor basin depth of 18 meters with a 600-meter turning circle.

4. Factors of Port Competitiveness and Development

In this section, we compile a list of port characteristics that past studies have found to be significant in influencing port traffic performance as well as the various measures of port performances.

4.1. Port location

The location aspects of ports have received considerable attention. Before the introduction and development of containerization of highly efficient intermodal transportation system, a port's site clearly defined its hinterland.[m] [Weigend, 1958] suggested that an ideal site for a port should have sufficient space for its operation, easy entrance, deep water, a small tidal range and a climate that will not hamper port operations at any time of the year.

The roles of a port can be distinguished as that of a feeder port, regional or global hub port. A study by [Fleming and Hayuth, 1994] identified centrality, intermediacy and proximity as three key location attributes that confer competitive advantage to some ports and allow them to become hub. Relative to feeder ports, [Sutcliffe and Ratcliffe, 1995] pointed out that a successful (transshipment) hub port must be situated at a location where there is a minimum diversion from the main shipping lanes for the line haul vessels and distance to the markets served is short. In addition, [Cargo System, 1998] suggested a port that is ideally located with respect to the main axial truck routes and either the rich hinterland or the feeder connections, and supported with appropriate services is more likely to be chosen by carriers to be their hubs.

Location also affects the way a port should compete. Noting that ports in small island economies may be at a disadvantage compared to ports that are natural gateways to rich hinterlands, [Robinson, 2002] advocated that the former should position themselves in such ways to achieve cost leadership (economies of scale) or service differentiation (economies of scope) in order to attain growth in the fast changing and highly competitive environment. Meanwhile, some other researchers at that time reported the key success factors for the positioning of small-island seaports to achieve competitive advantage. By conducting two case studies with Bahamas Freeport and Malta Freeport, their findings suggested that small-island ports with no direct hinterlands must first focus on cost leadership and then develop value-added services after cargo has been attracted.

While ports with locations that are natural gateways to rich hinterlands are evidently in a better position to develop the sea-to-land interface and inland transport services, what was once a secure area[n] for a port to draw traffic from is no longer the case with the advent of double-stacking of containers on rail-cars and the establishment of inland intermodal hubs. Containers can now be shipped long distances across continents to make

---

[m]Ports situated at good geographical locations benefit from the advantages of a large local market and an opportunity to capture transshipment cargo at the intersection of major sea routes.
[n]This area is determined based on land distance of the area from the port.

connections with ports. Moreover, as larger and larger ships operating in alliances seek economies of scale, fewer ports will be served directly by the larger transoceanic vessels. These load-centered ports are able to send or draw their traffic far and wide, by water mode or rail or truck, thus expanding the hinterland of the port. Consequently, ports are not just in competition with ports in their local area and along their immediate seaboard, they also are in competition with ports on distant seaboards attempting to serve the same inland areas [Haynes et al., 1997] [Fleming and Baird, 1999].

As competitions among distant seaports intensify, there is a growing role of intermediacy in a port's traffic structure that contributes to shrinking captive hinterlands. For load centers that depend on their intermediacy for cargo traffic, their positions are highly dependent on the ever changing service networks of shipping lines service network. In view of this, [Heaver et al., 2001] asserted that global carriers and forwarders challenge a port's capacity to influence goods flow and that ports can no longer expect to attract cargo simply because that are natural gateways to rich hinterlands. This point had also been noted earlier on by [Slack et al., 1985] and subsequently by [Carbone and De Martino, 2003].

With regards to the challenge from peripheral ports, spatial expansion on new sites may be brought about by two seemingly different factors. Obsolescence of older facilities ties is behind the Bird's 'Anyport'–based explanations. [Bird, 1963 and 1971] postulated that a shift in activity comes about as a result of a search for new sites that could offer space for mechanized terminal operations and/or sites adjacent to deepwater channel that would allow access for ever larger ships. [Frankel, 1987] and [Baird, 1996] supported Bird in their argument that the deeper waters[o] and the ocean locations of downstream ports allow the shipping lines to deploy their largest vessels and save sailing times and improve port turnaround time. Compared to river or upstream ports, ships can avoid the need for transiting through the long, narrow, inland waterways. The [Hayuth, 1981] model,[P] on the other hand, implies the congestion and diseconomies at

---

[o][Notteboom et al., 1997] disputed the argument, noting that the increase in vessel capacity is accounted more by an increase in the beam of ships rather than a draught. The authors cited the continued success of Hamburg and Antwerp as evidence for the persistence of upstream ports.

[P]In the light of Hayuth's five-stage load-center model, [Wang, 1998] examined the development of the Hong Kong container port in a regional context. The port-hinterland relationship between Hong Kong and China is found to be unique as the hub. Later,

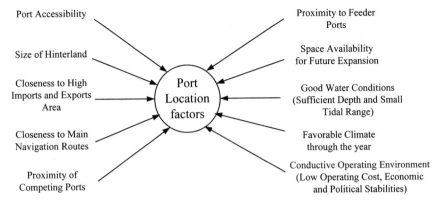

Fig. 7   Location factors affecting port attractiveness and competitions.

established terminals as the causes of competition. Other factors that have been put forward include access to shipping lanes, the search for deep-water sites, labor cost differentials, and environmental restrictions. According to Hayuth, factors that are important for the development of a load center port are the large-scale local market, high accessibility to inland markets, advantageous site and location, early adoption of new technological and social systems, aggressiveness of port management and the economic and political incentives of ports. The geographical location factors affecting port competition may be summarized in Fig. 7.

4.2. *Port efficiency*

Port efficiency is an important determinant of shipping costs, especially with the ever increasing vessel sizes that increase the unproductive cost of vessels waiting for services at ports. According to [Voorde, 2005], improving port efficiency from the 25th to the 75th percentile can reduce shipping costs by 12 percent. Besides, inefficient ports may equal to be 60 percent farther away from markets for the average country.

Since its inception, containerization has gained popularity and has become an essential component of a unit-load-concept in international sea

---

[Slack and Wang, 2002] examined the concept of the peripheral port challenge from an Asian perspective by focusing on the local and regional competition faced by Ports of Hong Kong, Singapore and Shanghai. The authors confirmed that these ports are subjected to challenges from peripheral ports in accordance with established models of port development but also suggested that the causes go beyond the challenge those postulated by the Hayuth's models.

freight transportation today. Hence, many researchers have examined the issue of achieving port efficiency via operations optimization in the context of container terminals. Among these studies, the adequate provision and effective allocation of berthing facilities are seen to have received much attention due to the fact that berthing (including waiting time of ships to load and unload their cargo) accounts for a significant portion of vessel time at port and inefficient berthing entails unnecessary productivity loss. [Steenken *et al.*, 2004] provided a detailed container terminal management and optimization review from more than 200 papers. His review described and classified the operations process in container terminals, and presented a survey of methods employed in the optimizations of quayside transport, the landside transport and crane transport, as well as, storage and staking logistics. [Vis *et al.*, 2003] also presented an overview survey on transshipment of containers at a container terminal, organized in accordance to the processes at container terminals: arrival of the ship, uploading and loading of the ship, transport of containers from ship to stack and vice versa, stacking of containers and inter-terminal transport and other modes of transportation.

Apart from the specifics in managing operations, it should be noted that the provision of up-to-date facilities, equipments and information technology (IT) infrastructures play an important leading role in enhancing the overall port efficiency. Nonetheless, investment in modern equipments to increase capacity is more complicated in practice as [Song, 2002] demonstrated the value of intelligent facilities investment in a port's success. Figure 8 is a summary of significant infrastructural factors affecting port operations efficiency.

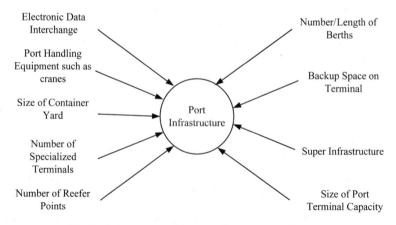

Fig. 8   Infrastructural factors affecting port efficiency.

4.3. *Multimodal network*

Ports often exist as part of a multimodal network, where various modes of transport (road, rail, water) are utilized in the successive movements of goods in an identical loading unit without any handling of the goods themselves during the transfers between modes. [Fleming and Hayuth, 1994], [Haynes *et al.*, 1997], [Klink and van den Berg, 1998] and [Helling and Poister, 2000] suggested that the ability of the port to support intermodal services could be a major factor in the container liners' choice of port of call. As such, port choice becomes more a function of network costs. The ports that are being chosen are those that will help to minimize the sum of sea, port and inland costs. By supplying intermodal services, ports can also open new markets beyond their traditional hinterland and create a competitive advantage over other ports without access to intermodal system.

The intermodal (or multimodal) networks in EU and US have been explored in [Caramia and Guerriero, 2009]. However, these models developed in the contexts of the EU or US may not effective in Asia because the regional patterns of market concentrations and/or logistic chains in Asia are fundamentally different from EU and US. In North America, markets are concentrated in east and west coast districts. These, the coastal cities, which are port cities at the same time, are connected by highway. In Western Europe, the Europe continent forms the hinterland, where markets are concentrated and linked to ports cities around the boundaries. In Southeast Asia, the concentrated markets are located at the port cities which are separated from one another.

Notwithstanding the unique characteristics of the EU, US and Asia continents, network optimization that has far-reaching implications on the development of ports remains as a substantial problem for any large scale of intermodal transportations. Broadly speaking, network operations take into account of infrastructure planning, service schedules, routing and pricing of services, location of international intermodal terminals, and their associated daily operations. Because of the inherent complexity that arises in international intermodal transportation, most of these network operation issues have been explored in the perspective of uni-modal transportation. Putting peripheral concerns aside, in comparison to local uni-modal transportation, international intermodal transportation faces a most strategic problem that is frequently characterized by (i) multiple (or conflicting) objectives such as minimization of cost and/or transport time and maximization transport capability; (ii) scheduled transportation modes; and (iii) time window constrains. Figure 9 is the multimodal

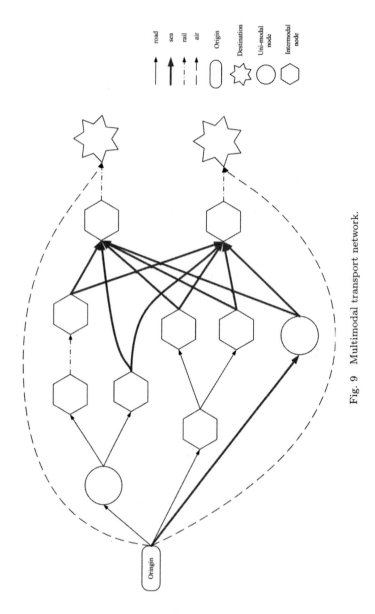

Fig. 9   Multimodal transport network.

transport network of one origin, two destinations and four transport modes, constructed in the contexts of South and Southeast Asia.

Literature on the route optimization of intermodal network is relatively sparse. Based on the international intermodal transportation in Asia-Pacific region, [Min, 1990] developed a goal programming model with chance-constraint to choose the most effective intermodal route which minimizes cost and risk while satisfying various on-time service requirements. The author tested this model by optimizing the intermodal routes between suppliers in Japan and manufactures in New York. Later, [Bookbinder and Fox, 1998] contributed an intermodal routing study for Canada-Mexico shipments under North American Free Trade Agreement (NAFTA). In their research, a network is constructed between five Canadian origins and three Mexican destinations. A shortest algorithm is proposed to calculate routes which minimize cost and throughout time. Most recently, [Chang, 2008] formulated a multi-objective, multimodal and multi-commodity flow problem with time windows and concave costs. The author optimized the route from three LCD suppliers in Taiwan to PC manufacture located in Denver, USA.

For the regional intermodal transportation network problems, [Modesti and Scimachen, 1998] presented a model to find shortest paths between various origin and destination pairs within the urban intermodal transportation networks of Genoa. Specific to the regional maritime transportation network, [Al-Khayyal and Hwang, 2007] formulated a model for finding a minimum cost routing in a network for ships engaged in pickup and delivery of liquid products. Their research analyzed the trade-off between transport costs and inventory costs in maritime routing across the whole supply chain within the Korean maritime networks. For the regional intermodal network of rail and road, [Caramia and Guerriero, 2009] studied a long-haul freight transportation problem with multiple objective functions and a focus on the implementation of a service network design that best to satisfy specific customer requests.

In a nutshell, the two major factors affecting multimodal network are (1) proximity/integration to trucking, rail and air transport; (2) competitiveness of complementary and substitutable trucking, rail and air transport.

4.4. Maritime trade strategy and institutional settings

The existing economic foundation of maritime and supporting industries, to a fair extent, determines the degree of aggressiveness in the pursuit of expansionary maritime trade strategies. At the same time, maritime trade strategies set the direction and lay the economic foundation for the developments of port and maritime industry and the nation as a whole. The effect of the undertaken maritime trade strategy may be reflected on the export production price and import price, import and export volume, open or protective trade mode.

[Francois and Wooton, 2001] proposed several maritime strategy models for different market structures to examine the implications of liberalization for profits, trade, and national gain from maritime trade. Using data from South Asia, Latin America and Southern Africa, Francois and Wooton confirmed that trade liberalization can lead shipping firms to capture a significant proportion of the benefits in multilateral trade concessions, and General Agreement on Trade in Service (GATS) negotiations in this area have important implication for multilateral efforts aimed more broadly at trade liberalization. By means of theoretical explicative and simulation modeling, [Coto-Millan et al., 2005] determined national and world income, prices of imports, exports and maritime transport services, and the utilizations degree of the productive capacity as the key variables that explain the behavior of maritime imports and exports for a particular economy. Quite recently, [Li and Cheng, 2007] collected data from 30 maritime WTO-membership nations and concluded that maritime policy is firmly based on economic conditions rather than the result of a rational analysis of policy makers in their relationship exploratory research.

An appropriate maritime strategy devised under considerations of the institutional settings will aid port development. As [Loo and Hook, 2002] commented, the evolution and competitive position of a container port needs to be understood as the interplay of international, national and local factors. [Fung, 2001] attempted to provide a systematic treatment for the interactions between the ports of Singapore and Hong Kong, and to investigate how the rise of South China ports affects the demand for Hong Kong container handling services. By including the Shenzhen port's throughput volume and various external trade variables as exogenous variables, the study also allowed for sensitivity analysis of their possible effects on the conditional growth path. Later, [Cullinane et al., 2004] analyzed the port of Shenzhen using Robinson's criteria for hub port development in an attempt to discern whether it will take over the role of Hong Kong to become the dominant regional hub. Their study concluded that despite Shenzhen's current competitive advantages, Hong Kong would, in all probability, retain its dominant role owing to other institutional

Table 1   Port characteristics on performances.

Port efficiencies	Frequency of Port Calls
	Port Berthing Time Length
	Labor Problems
	Economies of Scale (e.g., Cargo Volume/Size of Port)
	Diseconomies of Scale (e.g., Cargo Crowding Out Effect/Port Congestions)
	Material Handling Efficiency
	Goods Loss and Damage
	Flexibility of Operations Process — Large/Odd size Freight, Large Volume Shipment, Special Handling
Port productivity	Rate of Container Movement
	Number of Cranes Moves per hour
Port service	Working (or Port Operations) Hours
	Shipment Information
	Provides Assistance in Claims Handling
	Offers Convenient Pickup and Delivery Times
	Port Service Coverage — Routes
	Reliability
Port management	Management Expertise and Aggressiveness
	Number of Port Operators
	Privilege Contracts to Shippers/Carriers
	Government Taxes and Incentives
	Bureaucracy, Custom Administration and Regulations
	Coordination Between Departments
Port-induced cost	Port Charges
	Handling Charges
	Loading/Discharging Rate
	Inland Freight Rates
Others	Port Reputation
	Port Security (Port Safety/Terminal Security)

concerns. Other than direct head-on competition, [Song, 2002 and 2003] examined the possibility of co-operation between adjacent container ports in Hong Kong and South China as another maritime strategy.

Section 1.4 has explored into the various factors such as port location, port efficiency, multimodal network and maritime strategy that may have a bearing on port development. Table 1 provides further descriptions of the port characteristics that promote port performances in various dimensions depicted in Table 2.

5. Lessons to be Learnt

Evolutions of maritime trade over the centuries have revealed that there exists a close relationship between the development of a maritime industry

Table 2   Measures of port performances.

Competitive performance	Volume of TEUs handled
	Market Share
	Annual Growth
Operational performance	Ship Turnaround time
	Pre-berthing waiting time
	Percentage of idle time at berth to time at working berth
Asset performance	Output per ship berth day
	Berth throughput rate
	Berth occupancy rate
	Idle rate of equipment
Financial performance	Operating Surplus per ton of cargo handled
	Rate of return on turnover
	Proportion of General Cargo to Total tonnage handled

and its environment. Every emergent of a hub port city or prosperous maritime trade region is seen to astutely leverage on its own advantages, which may be a strategic location in proximity to major trading axes and affluent markets, advanced transportation system, superior services and/or modern management practices, and catches on the bandwagon during the special bull period. In turn, a budding maritime industry spurs port developments just as the opening of Suez Canal promotes the growths of many ports in Italy, Greece, India and Malacca Strait along the Asia-Europe line. The pace and degree of port development can be further facilitated by intelligent hard and soft investment in port infrastructure and management, made under careful considerations of its intrinsic characteristic relative to its peers.

In this section, we present some fruits for thought related to opportunities and threats for future port and maritime development in Asia. First of all, the associations between maritime trade and its business environment will irrevocably continue to hold. With the progressive global integration process, world trade is expected to expand and increase the demand for cargo shipping. More reductions of tariff in maritime trade will be reached through forthcoming conference agreements with lower trade barriers resulting to benefit both the producers and customers. As trade liberalizations across regions continue and invigorate their district economies, more long haul and short sea shipping will be induced and subsequently be developed into highway transportation on sea. At the meantime, China's Open Door Policy and its membership of the World Trade Organization (WTO) will provide further opportunities for ports that are based in China's manufacture centers (such as Shanghai, Shenzhen and Tianjin), and also for Singapore and Hong Kong ports that are situated on the crucial route providing China its needed energy and resources.

Second, maritime trade is a part of world economy which directly reflects the sentiments of exports and imports conditions. It has been observed that there is an obvious pronounced change in maritime trade industry after every major downturn or upturn of economy. During the economy recession, consuming markets will shrink, and the volume of world's import and export dramatically decreases. Contrarily, every boom of economy brings along a boom in maritime trade, as seen from the dramatic developments of both maritime trade and economy in Japan, Korea, Singapore and Taiwan in 1970s. Therefore, in addition to the specifics of port investment, prevailing and forecasted future economic conditions would be another element to be considered when deciding on the timing such investments.

Thirdly, mergers and partnerships in the liner shipping companies have allowed the deployment of ever larger and fewer vessels, with a primary motivation to reduce operations cost through economies of scale. However, this concentration within the liner shipping industry has increased the potent impact of a move by a major port user on the port's traffic. Coupled with advances in logistical systems that lead to the overlapping hinterlands of ports, the port industry faces heighten competition that entails the need for greater efficiency in ports (particularly, landside technological improvement will be required to reduce the waste or relieve bottlenecks at landside operations) and an ability to meet the changing demands of liner shipping companies. Nonetheless, [Low *et al.*, 2009] noted that the hub-and-spoke systems of operations in the liner shipping companies has also opened up co-operation opportunities for port to engage in collaborative efforts. Besides the port and maritime industry, the economy at large will also benefit from such scale increases and collaborative efforts in the process.

Fourthly, with the advancements of communication and transportation technologies that transcend time and space, the business competition will be in the global market in the coming decades. Despite the intensification of competition, more benefits can expected to be reaped from the scale of global industry that includes product design, manufacturing, orders processing, transportation, retailing and other miscellaneous service sectors. At the same time, greater emphasis will be placed on maritime trade strategy that optimizes the routing decisions of the entire value chain, spanning across geographical boundaries and comprising multiple transport modes. Naturally, ports as a component of the logistics chain will also have a more significant role to play a key role in export competitiveness, exports and import prices, as well as, the overall competitiveness of an economy.

Last but not least, port city development may cause land-use conflic-
tion, environment pollution and transportation congestion. We see that
unrestrained investment may not be a good strategy as seen in the case
of the Japanese ports, where there are insufficient traffic and excessive
capacity. Instead of being overly port centered, a lot of other factors
will need to be considered as well. For example, Hong Kong, having
a high population density, high labor cost, limited inland space and
traffic capacity, may face problems related to (noise and air) pollutions,
congestions and financial impediments in its future development. In Japan,
private transport is very expensive and public transportation is often too
crowded. Nevertheless, the impact of the latter in terms of pollution and
traffic congestion is less severe on a per head basis. In Korea, traffic
congestions have resulted in long hours of jams during peak period. For
the developing countries like India, [Pacione, 2006] highlighted problems
such as the inadequacy of infrastructure (i.e., safe drinking water, hygienic
sewage and low cost housing), inability to handle natural disasters (i.e.,
flooding) and control environmental pollution (gaseous emissions). It may
be fortunate for China, who has benefited from the experiences of these
developed countries, to construct an artificial port totally on the sea far
away from the Shanghai city. Such decision not only helps to alleviate
problems related to congestions and pollution, it also effectively solves the
problem that the water near Shanghai city is not deep enough for large
vessels. Another alternative to reduce environmental pollution in the port
city is the development of dry port. The dry port concept is based on a
seaport directly connected by rail with inland intermodal terminals where
shippers can leave and/or collect their goods in intermodal loading units as
if directly at the seaport [Roso, 2007].

6. Conclusions

This chapter tracks maritime trade evolution in Asia from the thirteenth
centuries to the post-World War II, with an analysis of the recent
developments in some of the Asian major ports. Through an extensive
review of the extant literature, this chapter identifies factors that have been
found to affect the port's standings in various dimensions of operating and
financial performances. Particularly, the impact of port location, inter-port
competition, multimodal network, maritime strategy and institutional settings on how a port should compete are discussed in depth with references to
the classic and contemporary theories. Additionally, the chapter highlights

some possible opportunities and threats facing the future developments of the port and maritime industry in Asia. We conclude the chapter with our beliefs that growths in the Asia's economy and its maritime trade are promising.

References

1. Al-Khayyal, F. and Hwang, S. (2007). Inventory constrained maritime routing and scheduling for multi-commodity liquid bulk, Part I: Applications and model, *European Journal of Operational Research*, 176, pp. 106–130.
2. Bird, J. (1963). The major seaports of United Kingdom. (Hutchison, London).
3. Bird, J. (1971). Seaports and Seaport Terminals. (Hutchison, London).
4. Bong, M.J. (2009). Challenges and Opportunities of Port Logistics Industry in Korea. (Korea Maritime Institute).
5. Bookbinder, J.H. and Fox, N.S. (1998). Intermodal routing of Canada-Mexico shipments under NAFTA, *Transportation Research Part E*, 34, pp. 289–303.
6. Caramia, M. and Guerriero, F. (2009). A heuristic approach to long-haul freight transportation with multiple objective functions, *OMEGA*, 37, pp. 600–614.
7. Carbone, V. and De Martino, M. (2003). The changing role of ports in supply-chain management — an empirical analysis, *Maritime Policy and Management*, 30, pp. 305–320.
8. Cargo Systems (1998). Opportunities for Container Ports. (London IIR Publication).
9. Chang, T. (2008). Best routes selection in international intermodal networks, *Computers and Operations Research*, 35, pp. 2877–2891.
10. Chua, T., Ingrid, R.L., Gorre, S., Adrian Ross, Stella Regina Bermad, Bresilda Gervacio and M. Corazon Ebarvia (2000). The Malacca Straits, *Marine Pollution Bulletin*, 41, pp. 160–178.
11. Coto-Millán, P., Carrera-Gómez, G., Baños-Pino, J. and López de Sabando, V. (2005). Rate of return regulation: the case of Spanish Ports, *International Advances in Economic Research*, 11, pp. 191–200.
12. Cullinane, K., Fei, W.T. and Cullinane, S. (2004). Container terminal development in Mainland China and its impact on the competitiveness of the Port of Hong Kong, *Transport Reviews*, 24, pp. 33–56.
13. Cullinane, K., Yap, W.Y. and Lam, J.S.L. (2006). The port of Singapore and its governance structure, *Research in Transportation Economics*, 17, pp. 285–310.
14. De Monie, G. (1995). The problems faced by Indian Ports today, *Maritime Policy and Management*, 22, pp. 235–238.
15. De, P. and Ghosh, B. (2003). Causality between performance and traffic — an investigation with Indian Ports, *Maritime Policy and Management*, 30, pp. 5–27.

16. DeGaspari, J. (2005). Ports look outward: seaports, shippers, and the government are linking their efforts to maintain a critical infrastructure, *Mechanical Engineering*, 127, pp. 30–35.

17. Fleming, D.K. and Hayuth, Y. (1994). Spatial characteristics of transportation hubs: centrality and intermediacy, *Journal of Transport Geography*, 2, pp. 3–18.

18. Fleming, D.K. and Baird, A.J. (1999). Comment — some reflections on port competition in the United State and Western Europe, *Maritime Policy and Management*, 26, pp. 383–394.

19. Francois, J.F. and Wooton, I. (2001). Trade in international transport services: the role of competition, *Review of International Economics*, 9, pp. 249–261.

20. Frankel, E.G. (1998). China's maritime developments, *Maritime Policy and Management*, 25, pp. 235–249.

21. Fujita, M. and Mori, T. (1996). The role of ports in the making of major cities: Self-agglomerations and hub-effect, *Journal of Development Economics*, 49, pp. 93–120.

22. Fung, K.F. (2001). Competition between the Ports of Hong Kong and Singapore — a structural vector error correction model to forecast the demand for container handling service, *Maritime Policy and Management*, 28, pp. 3–22.

23. Grant, P. (2002). Ex oriente luxuria: Indian commodities and Roman experience, *Journal of the Economic and Social History of the Orient*, 45, pp. 40–95.

24. Haralambides, H.E. and Behrens, R. (2000). Port restructuring in a global economy: an Indian perspective, *International Journal of Transport Economics*, 27, pp. 19–39.

25. Harper, T. (2002). Globalism and pursuit of authenticity: the making of a diasporic public spere in Singapore, *Sojourn*, 12, pp. 261–292.

26. Haynes, K.E., Hsing, Y.M. and Stouch, R.R. (1997). Regional port dynamics in the global economy: the case of Kaohsiung, *Maritime Policy and Management*, 24, pp. 93–113.

27. Hayuth, Y. (1981). Containerisation and the load centre concept, *Economic Geography*, 57, pp. 160–176.

28. Heaver, T., Meersman, H. and van der Voorde, E. (2001). Cooperation and competition in international container transport: strategies for ports, *Maritime Policy and Management*, 28, pp. 293–305.

29. Helling, A. and Poister, T.H. (2000). U.S. maritime ports — trends, policy implications and research needs, *Economic Development Quarterly*, 14, pp. 298–315.

30. Imai, A., Nishimura, E. and Papadimitriou, S. (2001). The dynamic berth allocation problem for a container port, *Transportation Research Part B*, 35, pp. 401–417.

31. Iris, F.A. Vis and Rene, de Koster (2003). Transshipment of containers at a container terminal: an overview, *European Journal of Operational Research*, 147, pp. 1–16.

32. Irwin, D.A. and Tervio, M. (2002). Does trade raise income? evidence from the twentieth century, *Journal of International Economics*, 58, pp. 1–18.
33. Klink, H.A., Van and Berg, G.C. (1998). Gateways and Intermodalism, *Journal of Transport Geography*, 6, pp. 1–9.
34. Lam, J.S.L. and Yap, W.Y. (2006). A measurement and comparison of cost competitiveness of container ports in Southeast Asia, *Transportation*, 33, pp. 641–654.
35. Lee, S., Song, D. and Ducruet, C. (2008). A tale of Asia's world ports: the spatial evolution in global hub port cities, *Geoforum*, 39, pp. 372–385.
36. Li, K.X. and Cheng, L. (2007). The determinants of maritime policy, *Maritime Policy and Management*, 34, pp. 521–533.
37. Lin, M. (1998). Maritime contacts between China and Rome, The Western Regions and the Chinese Civilization during the Han and Tang, pp. 307–321.
38. Liu, X. (1988). Ancient India and Ancient China: Trade and Religious Exchanges, AD 1-600, (Oxford University).
39. Loo, B and Hook, B. (2002). Interplay of international, national and local factors in shaping container port development: a case study of Hong Kong, *Transport Reviews*, 22, pp. 219–245.
40. Low, J.M.W., Lam, S.W. and Tang, L.C. (2009). Assessment of hub status among Asian ports from a network perspective, *Transportation Research Part A*, 43, pp. 593–606.
41. Min, H. (1990). International intermodal choices via chance-constrained goal programming, *Transportation Research Part A*, 25, pp. 351–362.
42. Modesti, P. and Sciomachen, A. (1998). A utility measure for finding multi-objective shortest paths in urban multimodal transportation networks, *European Journal of Operational Research*, 111, pp. 495–508.
43. Notteboom, T.E. (1997). Concentration and load centre development in the European container port system, *Journal of Transport Geography*, 5, pp. 99–115.
44. Pacione, M. (2006). City profile — Mumbai, Cities, 23, pp. 229–238.
45. Robinson, R. (2002). Ports as elements in value-driven Chain systems — the new paradigm, *Maritime Policy and Management*, 29, pp. 241–255.
46. Roso, V. (2007). Evaluation of dry port concept from environmental perspective: a note, *Transportation Research Part D*, 12, pp. 523–527.
47. Sen, T. (2006). The formation of Chinese maritime networks to Southern Asia, 1200–1450, *Journal of the Economic and Social History of the Orient*, 49, pp. 421–453.
48. Slack, B. (1985). Containerization, inter-port competition and port selection, *Maritime Policy and Management*, 12, pp. 293–303.
49. Slack, B. and Wang, J. (2002). The challenge of peripheral ports: an Asian perspective, *GeoJournal*, 56, pp. 159–166.
50. Song, D.W. (2003). Port co-opetition in concept and practice, *Maritime Policy and Management*, 30, pp. 29–44.
51. Song, D.W. (2002). Regional container port competition and co-operation — the case of Hong Kong and South China, *Journal of Transport Geography*, 10, pp. 99–110.

52. Steenken, D., Vob, S. and Stahlbock, R. (2004). Container terminal operation and operations research — a classification and literature review, *OR Spectrum*, 26, pp. 3–49.

53. Sutcliffe, P. and Ratcliffe, B. (1995). The battle for med hub role, *Containerization International*, 28, pp. 95–99.

54. Tan, T. (2007). Port cities and hinterlands: a comparative study of Singapore and Calcutta, *Political Geography*, 26, pp. 851–865.

55. Tang, L.C., Low, J.M.W. and Lam, S.W. (2008). Understanding port choice behavior — a network perspective, Network and Spatial Economics, 10.1007/s11067-008-9081-8.

56. United Nations Conference on Trade and Development Secretariat (2008). Review of Maritime Transport 2008.

57. Voorde, E.E.M.V.D. (2005). What future the maritime sector? Some considerations on globalization, co-operation and market power, *Research in Transportation Economics*, 13, pp. 253–277.

58. Wang, J.J. (1998). A container load with a developing hinterland — the case of Hong Kong, *Journal of Transport Geography*, 6, pp. 187–201.

59. Weigend, G.G. (1958). Some elements in the study of port geography, *Geographical Review*, 48, pp. 185–200.

60. Yabe, Y. (1991). Major characteristics of urban waterfront redevelopment in Japan, *Marine Pollution Bulletin*, 23, pp. 397–401.

61. Zhu, J., Lean, H. and Ying, S. (2002). The third-party logistics services and globalization of manufacturing, *International Planning Studies*, 7, pp. 89–104.

CHAPTER 2

RECENT DEVELOPMENT OF MARITIME LOGISTICS

Loo Hay Lee, Ek Peng Chew, Lu Zhen, Chee Chun Gan, and Jijun Shao

Department of Industrial & Systems Engineering,
National University of Singapore
1 Engineering Drive 2, Singapore 117576
iseleelh@nus.edu.sg

The economic crisis from the end of 2007 to early 2010 has greatly affected the maritime industry. Some of the biggest problems include oversupply of tonnage, sudden drop in demand, and declining freight rates.

Global containerized trade enjoyed a year-on-year increase of 4.6 percent in 2008. The liner shipping market accounts for 16 percent of the world's trade goods loaded in terms of tons. One of the most significant trends in liner development is the increase in vessel size. There are notable increases in both the size of the largest vessel and the average vessel size. The productivity of most container lines, however, decreased in 2008, partly due to the various cost control measures adopted by carriers, such as slow steaming, to mitigate the effect of the crisis. Freight rates also suffered in 2008 and 2009. With demand recovering in 2010, freight rates have increased, and have even reached pre-crisis levels in some trade routes.

For container terminals around the world, global operators are still the market leaders and this trend is likely to continue. Container transshipment activities have enjoyed rapid growth during the 1990s, and the share of transshipment activities in total container volume has reached equilibrium from 2000 onwards. There were some notable improvements to port performance in 2008. The United Nations' Liner Shipping Connectivity Index 2009 has clearly shown the effect of the crisis in the industry.

1. Development Trends on Global Container Shipping

1.1. *Global economic condition and industry perspective*

Global container shipping has been growing in the past 20 years before the current economic crisis. Figure 1 shows the growth of maritime

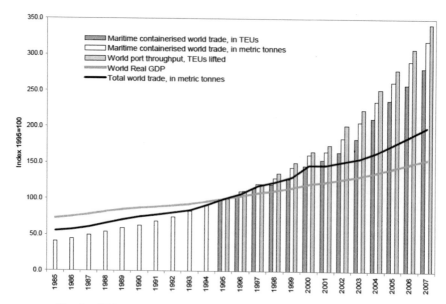

Fig. 1   GDP, trade and maritime containerised transport. 1985–2007.
Source: Sánchez, Ricardo J.[1]

containerized trade and world gross domestic product (GDP). The growth rate for seaborne containerized trade is higher than the growth rate of world GDP and total seaborne trade. There has been sustained growth even during years of past recessions.

The economic crisis of 2007–2010, also known as the Great Recession, has seriously damaged the global economy. In 2009, the world GDP declined by 1%, for the first time in the post-World War II era. The recession has also hit global trade, with trade volumes declining by as much as 25% in 2009 from 2008's level, the largest single year drop since World War II. Lower commodity prices, tighter credit, and increasing protectionism all contributed to the drop in trade demand during the recession.[2]

The container shipping industry has been badly hit by the recession. Throughout year 2009, container volumes have declined, liners and terminal operators have suffered from diminishing profits or even losses, new container ship ordering has almost halted, vessels have been laid up by ship owners, and freight rates and charter rates have reached record low levels. The crisis has quickly caused a shake up in the container shipping industry.

There are quite a few reasons for the quick spread of the downturn to the maritime industry. One of the major reasons is that the demand for maritime transport services, especially container shipping services, is derived from economic growth and trade volume. As the economic condition is bad during the recession, the demand for trade, as well as the need for maritime transport services to carry goods from manufacturers to customers, has diminished. Another side effect of such strong dependency is that the global seaborne trade and container shipping is likely to rebound and recover from the recession slower than other sectors.[3] Another factor that affects the industry is the ever increasing gap between supply and demand for container shipping. The recession has caused a sudden and unexpected drop in the demand side. However, most of the orders for new ships were placed during a period of rapid growth for both the container shipping industry and the demand for shipping services. As these orders are approaching their delivery date, the growth in the supply side of the container shipping industry will outpace the growth, if not contraction, in demand. In 2009, the total container fleet is expected to grow at 9.6 percent while the demand is expected to go down by 9.1 percent. The growth rates of supply and demand in global container shipping from 2000 to 2009 are shown in Fig. 2. We can see that before the current crisis, the growth rates of demand and supply closely matches each other. However, in 2009, there is a huge gap between the growth rates for demand and supply. The gap in the growth rates would likely lead to the gap in the actual figure

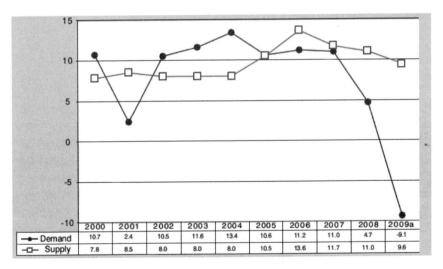

	2000	2001	2002	2003	2004	2005	2006	2007	2008	2009a
Demand	10.7	2.4	10.5	11.6	13.4	10.6	11.2	11.0	4.7	-9.1
Supply	7.8	8.5	8.0	8.0	8.0	10.5	13.6	11.7	11.0	9.6

Fig. 2 Growth rate of supply and demand in global container shipping.
Source: the UNCTAD secretariat[3]

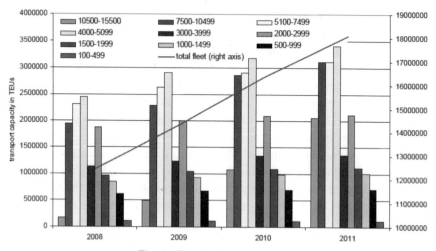

Fig. 3   Expected container fleet.

Source: Sánchez, Ricardo J.[1]

of demand and supply, which will likely cause huge overcapacity in the industry. Furthermore, as new vessels typically take years from point of order to delivery, the existing orders are likely to keep the growth rate high on the supply side for several years and thus enlarge the overcapacity problem until they are fully delivered in 2011. Figure 3 illustrates the expected total container fleet at the end of each year to 2011, based on current orders. We can see from the chart that the speed of the capacity expansion is quite fast, and the problem of overcapacity would likely to remain in years to come. This overcapacity problem has been recognized by many industry experts, and carriers have adopted several measures to deal with the problem. Orders are being cancelled or delayed by carriers, and excessive capacity is being removed by laying up idle ships and demolishing older vessels.

✛ Many industry experts feel that the unavailability of trade financing during the downturn has also made a significant contribution to the rapid spread of the downturn in the maritime industry. Since the recession started as a financial crisis and banks are taking extra caution on lending, it is much harder for carriers, shippers and port operators to obtain sufficient financing for projects. According to the UNCTAD Secretariat, the shortage of financing is most severe in developing economies, with unmet demand estimated to range between $100 billion and $300 billion annually.[3]

The recovery from the recession started at the end of 2009 and continued into the first half of 2010. The global economic condition has become better, and the recession has been considered to have ended in the first half of 2010. However, there is still substantial risk in the shipping industry in 2010. In a survey conducted in September 2009 by AMR Research, "the recovery cycle" has been identified as the biggest risk in 2010. The main reasons given in the survey include "potential commodity price increases, limited internal skills after work force reductions, and problems meeting new demand with constrained capacity, low inventory and transportation constraints."[4] Europe-based shipping industry analyst group Seabury has forecasted that global seaborne container trade will grow about 11 percent this year over 2009.[5] There is much uncertainty in the container shipping industry as well as the global economic condition during the recovery cycle in 2010. Many carriers remain cautious about the economic prospect, and some shipping executives still feel that the downturn has not brought enough needed consolidation and efficiency to the container shipping industry and overcapacity may still be a big problem during the recovery.[6]

1.2. *Recent trends in container shipping industry*

The world container ship fleet continues to expand in 2008 despite the economic crisis. Table 1 Development in World Container Ship Fleet shows the development in global container ship fleet for the past decade. We can see that there were 4,638 ships by the beginning of 2009, with a total capacity of 12.14 million TEUs. This represents an increase of 8.5 percent in the number of vessels from 2008 to 2009. In terms of TEU capacity, there is a 12.9 percent increase. Another interesting trend in the world container fleet is that the average vessel size increased from 2516 TEU in 2008 to 2618 TEU in 2009, representing a 4 percent growth. For new container ships, the

Table 1   Development in world container ship fleet.

World total	1987	1997	2007	2008	2009	Growth 2009/2008
Number of vessels	1 052	1 954	3 904	4 276	4 638	8.47
TEU capacity	1215 215	3 089 682	9 436 377	10 760 173	12 142 444	12.85
Average vessel size	1 155	1 581	2 417	2 516	2 618	4.04

Source: the UNCTAD secretariat[3]

trend towards larger vessels is much clearer. In 2008, the average size of new container ships entering service is 3489 TEU, an increase from 3291 TEU in 2007. On 31 October 2009, there were 218 new 2009-built fully cellular container ships in service with average carrying capacity of 4,125 TEU.[3]

Another trend in the world cellular fleet is the trend towards more gearless vessels. Among 2008-built container ships, nearly 80 percent of vessels and 90 percent of TEU capacity are gearless, comparing to only about half of the vessels built 10 years ago. This is also because of the development in port facilities, since more and more ports are equipped with modern handling equipment, especially specialized gantry cranes. The trend towards gearless vessels is likely to encourage ports to invest further in port handling equipment.[3]

2. Liner Shipping

2.1. *Container liners*

The container shipping market has grown rapidly over the past 20 years. In 2008, global containerized trade was estimated at 1.3 billion tons, which represents a year-on-year increase of 4.6 percent. The liner shipping market accounts for 16 percent of the world's trade goods loaded in terms of tons.[3]

Liner shipping companies are integral components of the global container shipping network, and their decisions and policies obviously have large effects on the liner industry. The past few years have been a difficult time for most liners and they have had to make many adjustments in order to stay afloat in the face of sharp reductions in volume and severe overcapacity.

2.1.1. *Liner market developments*

The decade from 2000 to 2010 saw further concentration of the container market in the hands of the top 10 liner operators. The capacity share of the top 10 operators grew from 49 percent to 58 percent. The majority of the increase can be accounted for by the growth from Maersk, MSC and CMA-CGM. Maersk's and CMA-CGM's growth arose mainly from mergers and acquisitions, whereas MSC's increase in capacity share came from organic growth.

As of 1 January 2010, AXS-Alphaliner's Top 100 Report on the fleet capacity share of the top 20 liner operators around the world includes 13 liners from Asia, reinforcing Asia's position as the leading region for containerized trade. However, several Asian carriers have also significantly reduced the size of their container fleet over the past year. Most notably, the three Japanese carriers NYK, MOL and "K" Line have decided to reduce their exposure to the volatile liner trade as part of a long term corporate strategy shift. According to AXS Alphaliner, seven major Asian operators have disposed of a combined 165 vessels totalling 282,000 TEU in the past 15 months. This includes 155,000 TEU sent for scrap and 127,000 TEU in secondhand sales.

Maersk Line maintained its dominance as the market leader in Jan 2010, but lost ground to second-ranked MSC in market share in terms of fleet capacity. Maersk's share dropped from 16 percent in 2007 to 15 percent in 2010 while MSC's share increased from 10 to 11 percent in the same period. However, almost all of the top 20 liner companies experienced massive reductions in their profits, with Hapag-Lloyd being the only exception.

2.1.2. *Increase in liner ship size*

The size of container ships has been increasing steadily over the past three decades, as shown in Fig. 4. In 1975, the largest container ship had a capacity of 3,000 TEU. In 1991, that increased to 4,000 TEU and 6,800 TEU in early 2000. Most recently in 2009, container ships with capacity of up to 14,000 TEU have been pushed into service. The vast majority of these ships are used on the Asia-North America and the Asia-Europe routes, which offer the optimum combination of high volumes, long voyages and deep, efficient ports. The construction of wider locks at the Panama

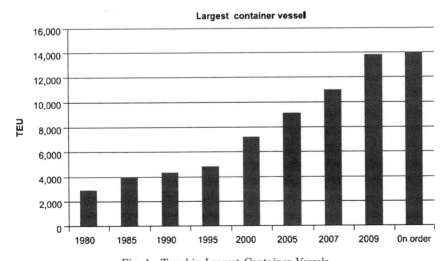

Fig. 4   Trend in Largest Container Vessels.
Source: Historical series compiled from Containerisation International, various years

Canal, which is expected to be complete by 2014, will provide carriers with additional options for deploying their super-sized container ships. The wider locks will be able to accommodate "New Panamax" ships of about 12,500 TEU, compared with the current Panamax capacity of about 5,000 TEU.

High-volume container ships can provide significant cost savings to liner companies because of the economies of scale in major trade lanes. A 12,000-TEU ship can be operated with the same 13 or 14 crewmembers required by a ship with half the capacity. Per-unit costs of capital investment and fuel consumption are substantially less than for two vessels of half the size. It is thus no surprise that large container ships also dominate the order books, with ships of capacity over 7,000 TEU making up 58 percent of the current capacity on order. 30 percent of the order book is comprised of 128 vessels of 12,000 TEU and above. In combination with the growth in the number of container ships worldwide, this trend represents a very significant increase in global container shipping capacity.

The increase in size of container ships has also lead to more hub-and-spoke structures in the global shipping network. This is because of the fact that only several major trade lanes have sufficient demand to support these mega vessels. Furthermore, the mega vessels can only dock in a small number of ports because of their deep drafts. It is thus natural for carriers to adopt a hub-and-spoke network structure in order to take advantage of the economies of scale by consolidating the demand from smaller trade lanes and ports. Consequently, the demand for transshipment operation has also increased due to the shift towards larger vessels.

Not all liner companies, however, are convinced that the trend towards increasing container ship sizes is healthy for the market. Evergreen has decided not to include any of these mega-containerships in their recent order for 100 new container vessels. Evergreen's order enables them to take advantage of the drop in newbuilding prices amidst the current oversupply in shipping capacity, but Evergreen chairman Chang Yung-Fa had previously expressed reservations on the market impact of such high-volume vessels, which are unable to take advantage of their economies of scale unless they are constantly full. With the current depressed global economy, it remains to be seen whether the trend of liners increasing their ship sizes will continue.

2.1.3. *Liner productivity*

Due to the high cost of assets such as container ships, liner companies are naturally concerned about liner productivity, which measures the utilization of such expensive assets. Unfortunately, the productivity of the world's shipping fleet in terms of ton-miles per deadweight-ton (dwt) has been decreasing since 2005 and is expected to decline further. By 2008, liner productivity had fallen to levels only 6 percent higher than those in 1990.

The decrease is due to a combination of many factors, most notably the oversupply of tonnage coupled with the decrease in trade growth. In 2009, the size of the world's container fleet was estimated to grow by 9.6% in contrast with a drop in demand of 9.1%. The cargo volumes carried by the world residual fleet, including container ships and general cargo carriers, dropped from 10.84 to 10.4 tons per dwt.[3]

Furthermore, liner companies have introduced various cost saving measures which further decrease liner productivity. Previously, liner companies introduced slow steaming in order to reduce fuel costs. With the recent decline in fuel prices, liner companies have also taken to choosing longer but cheaper routes, which has also resulted in fewer ton-miles per dwt. On the other hand, these longer routes, e.g. bypassing the Suez Canal via the Cape of Good Hope, can result in significant cost savings (up to $300,000 in some cases for the largest ship) even after considering additional fuel and crew costs, mainly due to savings on canal transit fees. Such rerouting also bypasses the hotspot for piracy near Somalia, which presents additional cost savings for insurance. It is also a good measure to absorb additional capacity, since such rerouting will increase the average sailing time by approximately 7 days. However, the viability of the reroute strategy relies on the current relatively low bunker prices. The long-term applicability of such strategies could come into question when bunker prices increase.

2.2. *Freight rates*

Liner companies have been hit hard by the sharp decline in freight rates since the start of the economic crisis. Freight rates for most vessels plummeted due to the decrease in shipping volumes, with April 2009 rates dipping below the levels experienced in 2000. Container volumes on the Asia-Europe route fell by around 15 percent in 2008. This contributed to a severe drop in freight rates in early 2009 to about $300 per TEU, a decline of 80% from the peak in 2007.[3]

Furthermore, in 2008 the European Union repealed the block exemption that had previously been granted to liner conferences with regards to price and capacity setting. This resulted in the former members of the Far East Freight Conference having to set their own tariffs and surcharges since 18 October 2008, presenting an additional burden to shippers in having to keep track of the multitude of different rates. The first quarter of 2009 saw a corresponding decrease in trade volumes on major routes to and from Europe. The Europe-Asia route declined by 22 percent on the westbound route from Asia, and the eastbound route from Asia experienced a 17 percent decline at the same time. The transatlantic westbound route to North America also experienced a 17 percent decline, while the eastbound route from North America to Europe dropped by 30 percent.

However, great improvements in freight rates can be seen recently due to recovering demand with the peak season and great efforts from liners in maintaining freight rate increases despite heavy competition. Transpacific freight volumes have increased 13% year-on-year in the first quarter of 2010, with corresponding freight rate increases to $2,607 per FEU compared to pre-crisis levels of $2,000 per FEU.

Such recovery does not come without cost, however. Shippers are becoming increasingly unhappy about sharp increases in rates, in addition to service changes such as slow-steaming which increase shipping time. Furthermore, it is unclear if the recovery in shipping volumes is sustainable rather than simply a case of retailers restocking inventory. A deterioration in volumes, combined with an expected wave of new containership deliveries (410,000 TEU in Q2 2010), may undermine the recovery potential in freight rates. Some shippers have proposed holding regular pricing discussions with liners in an effort to prevent seeing such sharp swings in freight rates, which also add uncertainty to shippers' supply costs. Liner companies, on the other hand, have also expressed concerns about regular communication and cooperation with shippers to reduce costs and conflicts on both sides.

3. Ports

3.1. *Global container terminal operators*

Global container terminal operators are companies with activities in more than one geographical region. Most global terminal operators are private organizations, while PSA, DP World, HHLA, COSCO Pacific and China Shipping are majority owned by elements attached to a state or public entity. Global container terminal operators have the highest market share in South East Asia and North Europe, controlling 76.6% and 75.5% of throughput respectively in these two regions. Global operators are least significant in Eastern Europe (30.3%), South America (32.2%) and Australasia (32.8%). Global operator throughput capacity is expected to rise slightly above the average by 2.7% annually between 2008 and 2014. By 2014, the combined market share controlled by global operators as measured by capacity will be 55.7%.

According to Drewry, global container terminal operators can be classified into three categories:

(1) Global stevedores: these are companies whose core business is terminal operation. They view terminals as profit centers. As a result, global stevedores often focus on achieving greater efficiency by implementing common systems across a terminal network. HPH, PSA and DP World are the leading stevedore-based global operators. This type of terminal operator accounts for 57.2% of the total throughput by global terminal operators in 2008.

(2) Global carriers: these are companies for which container shipping is the prime focus, and whose terminal networks support shipping activity. Terminals are often run as cost centers and efficiency gains are achieved through the integration of the terminals with a global shipping service network. Evergreen, Hanjin, K Line, China Shipping and MSC are the leading carrier-based global operators. This type of terminal operator accounts for 12.2% of the total throughput by global terminal operators in 2008.

(3) Global hybrids: these are companies focusing on both shipping and terminal operation. The main activity of such companies, or that of their parent groups, is container shipping, but separate terminal-operating divisions have been established. These terminal-operating divisions are expected to handle a significant amount of third-party traffic besides serving the core liner shipping business of the parent companies. They are often designed to operate as independent profit centers. For example, APM is the biggest and longest-established hybrid terminal operator. CMA, COSCO and APL/NOL are also the examples of this type of operator; their terminal operating divisions are Terminal Link, COSCO Pacific and APL Terminals respectively. This type of terminal operator accounts for 30.5% of the total throughput by global terminal operators in 2008.

There are some big stevedore-based terminal operators which are not considered as global terminal operators as they only focus on one country. The Shanghai International Ports Group and Gulftainer are the examples of these operators. They have expressed interest in expansion overseas. Thus more companies will be added to the existing list of global terminal operators in the future.

Some liners may step up investments in their sectors of terminal operations. For example, Hamburg Sud is reportedly participating in a terminal development in Santa Catarina, Brazil; UASC has taken a shareholding in a new development in the Egyptian port of Damietta. However, the uncertainty involved in current business climate may imply that such plans are put on hold for the immediate future.[7]

The financial crisis is likely to seriously curtail the investments in terminals in the short to medium term. Some carriers may be forced to dispose of their terminal assets. Bankruptcies of carriers and mergers and acquisitions between carriers are anticipated. It may create some larger liner-based global terminal operators over the next few years. In the first half of 2010, as the economic conditions improve, operators are once again considering future investment opportunities.

3.2. Leading terminal operators

Drewry ranks container terminal operators by two different forms of measurement:

Measurement 1: the summation of total throughput of all terminals in which the global operator has a stake of more than 10%.

Measurement 2: the weighted summation of terminal throughput based on the equity share of the operator in the terminal, which could correctly reflect the amount of interest of the operator in the terminal and eliminate double counting on terminals with multiple interests.

The league table of global operators according to the above two forms of measurement are listed as follows in Table 2:

The big four terminal operators (PSA, HPH, APMT, DPW) are currently the dominant players in the global container terminal sector, by both forms of measurement. There are quite a few challengers, like Cosco, MSC, and perhaps China's SIPG (Shanghai International Port Group), but their positions appear to be secure for the foreseeable future. By Measurement 1, which is the total throughput measure, these top four market leaders enjoyed a 45.6% share of world throughput in 2008. This compares with 45.1% in 2007, underlining the fact that the influence of these four companies within the container terminal market continues to strengthen. The fastest-growing operator by Measurement 1 is the Marseilles-based shipping group CMA CGM, which jumped from 15th (in 2007) to 11th (in 2008). The biggest rate of decline among the global operators was APL, which dropped from

Table 2   The league table of global operators' throughput in 2008.

Measurement 1				Measurement 2			
Rank	Operator	Million TEU	% Share	Rank	Operator	Million TEU	% Share
1	HPH	67.6	13.0	1	PSA	50.4	9.6
2	APMT	64.4	12.3	2	HPH	34.4	6.6
3	PSA	59.7	11.4	3	APMT	33.8	6.5
4	DPW	46.2	8.9	4	DPW	32.9	6.3
5	Cosco	32.0	6.1	5	Cosco	11.1	2.1
6	MSC	16.2	3.1	6	Evergreen	8.9	1.7
7	Eurogate	13.2	2.5	7	MSC	7.9	1.5
8	Evergreen	10.3	2.0	8	Eurogate	7.4	1.4
9	HHLA	7.4	1.4	9	HHLA	6.7	1.3
10	SSA Marine	7.4	1.4	10	SSA Marine	4.6	0.9
11	CMA-CGM	7.0	1.3	11	APL	4.2	0.8
12	Hanjin	5.7	1.1	12	CMA-CGM	4.1	0.8
13	Dragados	5.5	1.1	13	NYK Line	3.8	0.7
14	NYK Line	5.5	1.1	14	ICTSI	3.7	0.7
15	APL	5.4	1.0	15	Dragados	3.7	0.7
16	OOCL	3.9	0.7	16	Hanjin	3.6	0.7
17	ICTSL	3.8	0.7	17	K Line	2.6	0.5
18	K Line	3.4	0.6	18	MOL	2.4	0.5
19	MOL	3.2	0.6	19	GrupTCB	2.4	0.5
20	Grup TCB	3.2	0.6	20	OOCL	2.1	0.4
21	Yang Ming	2.0	0.4	21	Yang Ming	1.3	0.3
22	Hyundai	1.1	0.2	22	Hyundai	1.2	0.2
Total		374.1	71.6			233.1	44.6

Source: Drewry Shipping Consultant[7]

11th (in 2007) to 15th (in 2008). PSA of Singapore is the leading global container terminal operator by a huge margin according to Measurement 2. Its 50.4million equity TEU throughput in 2008 is more than 16 million above the second place, i.e., HPH.

Table 3 places the operators in order of the total available capacity at all terminals in 2008.

The company with the biggest capacity at the end of 2008 was APM Terminals. HPH and PSA also achieved solid increases in available capacity of around 5.4% and 7.6% respectively, allowing them to retain their second and third rankings in this capacity table. DP World and Cosco both achieved double-digit capacity growth in 2008. Among the top 20 global operators, 8 companies increased capacity by more than 10% between 2007 and 2008. The highest growth rates were recorded by OOCL, SSA Marine, NYK and ICTSI.

Table 3   The league table of global operators' capacity in 2008.

Rank	Operator	million TEU	% Change 07-08
1	APM Terminals	98.4	3.8
2	HPH	84.8	5.4
3	PSA	84.6	7.6
4	DP World	63.7	10.6
5	Cosco	61.9	13.1
6	MSC	22.2	13.2
7	Eurogate	20.7	16.4
8	Evergreen	16.3	1.2
9	SSA Marine	13.6	24.4
10	Hanjin	11.2	−0.6
11	NYK	10.2	22.0
12	CMA CGM	10.1	7.7
13	HHLA	9.8	4.9
14	Dragados	9.1	−9.0
15	APL	7.8	0.0
16	ICTSI	6.6	19.3
17	K Line	5.2	−10.3
18	OOCL	5.0	25.0
19	MOL	4.7	−6.9
20	Hyundai	4.3	2.7
21	TCB	4.2	−0.1
22	Yang Ming	3.6	−5.0

Source: Drewry Shipping Consultant[7]

Despite the uncertain short term prognosis for the world's container trades, the future outlook beyond this is still positive, with China, India and Vietnam in particular expected to push container traffic levels higher. Consequently, there is an underlying need for further expansion of container terminal capacity in the medium and long term. Indeed there are several regions where, once economic recovery is underway, container terminal bottlenecks could quite quickly strangle growth unless investment is sustained. Actually, during the first half of 2010, such trends have already been noted in some ports.

DP World is one the most acquisitive operators in recent years and continues this pattern in 2008. Notable developments include the purchase of a 90% shareholding in the operator of the Egyptian port of Sokhna, a 60% shareholding in a terminal operator in Tarragona, Spain, and so on. PSA invested in International Trade Logistics of Argentina and consequently obtained a stake in the Buenos Aires operator Exolgan. PSA also purchased substantial shareholdings in terminals in Kandla and Kolkata in India, and entered into a joint venture with PIL to develop and operate a terminal in Singapore.

PSA and HPH remain heavily dependent on traffic generated by terminals in the Far East and South East Asia. These two regions account for more than 81% of the PSA's equity TEU throughput, and 59% of HPH. Both PSA and HPH also have a significant presence in Europe, deriving respectively 16.8% and 25.5% of their equity TEU from European terminals. APM Terminals and DP World have the most geographically balanced spread of container terminal activities of all leading global container terminal operators.

The container terminal business is highly competitive and competition has further intensified since the economic downturn. The competition between global operators does exist within many of the world's leading container ports, including, for example, Busan, Kaohsiung, Hong Kong, Rotterdam, Antwerp, Hamburg, Long Beach and Los Angeles. In fact, of the world's top ten container ports, competition between global container terminal operator is absent only in Dubai, Singapore and China's Shanghai and Shenzhen ports.

Global container terminal operators are increasingly willing to cooperate where it is in their mutual interest to do so. A large number of the joint ventures in which two or more global operators have shareholdings are partnerships between a stevedore-based global operator and one or more carrier-based operators.

Last but the most important, the financial performance of the global container terminal operators is listed as follows:

As shown in Table 4, HPH and DP World achieved the highest turnover and earnings from their global container terminal operations. The third

Table 4   The financial performance of the selected global operators in 2008.

Terminal operator	Throughput	Turnover (million USD)	Earnings (million USD)
Eurogate	14.2	996.97	282.59
HPH	67.6	5109.08	1707.80
ICTSI	3.8	435.52	184.73
NYK Line	6.4	1447.78	65.81
PSA	63.1	3025.09	901.60
APMT	76.9	3119.00	573.00
Cosco Pacific	45.9	unknown	128.20
DPW	46.6	3283.00	1340.00
HHLA	7.4	1104.74	525.38
APL	5.4	577.00	93.00

Source: Drewry Shipping Consultant[7]

most profitable global operator is PSA. Among the top five global terminal operators, DP World achieved the best earnings per TEU, with $28.74 per TEU, followed by HPH with $25.25 per TEU, and PSA with $14.30 per TEU. APM Terminals and Cosco fared less well with earnings per total TEU of $7.45 and $2.79 respectively.

3.3. *Development in transshipment activities*

The container transshipment volume has been growing steadily. The proportion of transshipment volume in total container volume rose from 18 percent in 1990 to 25 percent in 2005.[8] The following figure illustrates the growth of the percentage share of transshipment activities in total port volume from 1990 to 2006.

The major driver of transshipment activities is the use of hub-and-spoke system by liners. The hub-and-spoke system can consolidate demand from small ports to hub ports, so as to achieve economies of scale. However, the hub-and-spoke system may also incur extra costs in longer transit time, extra handling at the hub port, etc. From Fig. 5 Growth in Transshipment activities (Source: Transport and Tourism Division in UNESCAP[8] we can see that the percentage share of transshipment activities increased rapidly during the 1990s, and at a much slower rate from 2000 onwards. The percentage share of transshipment activities has reached equilibrium, which

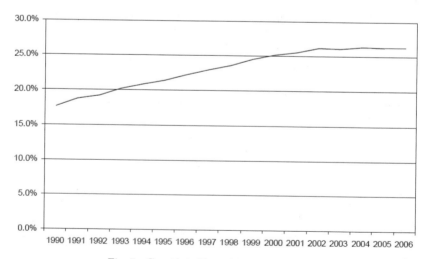

Fig. 5   Growth in Transshipment activities.
Source: Transport and Tourism Division in UNESCAP[8]

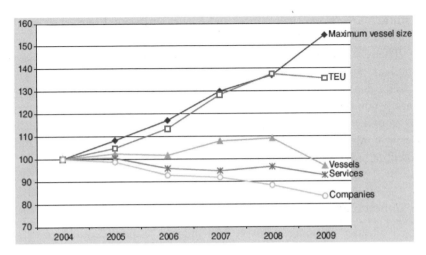

Fig. 6   Trends in LSCI Indicators.

is likely caused by stabilization in the hub-and-spoke network structure of major carriers.

According to UNESCAP estimates, global container transhipment volume will increase from around 85 million TEU in 2005 to 184 million TEU in 2015 at an average growth rate of 7.6 percent per annum. This growth rate is on par with the expected growth rate for total world container volume.

3.4. *Improvement in port performance*

Improvements on cargo handling in container terminals would greatly help the global supply chain. The most notable improvement to port performance in 2008, as mentioned by the Review of Maritime Transport 2009, is the increase in number of ports achieving higher crane productivity.[3] Following the trend of increasing vessel size, there is also much pressure on the efficiency of terminal handling operation. With the help of new training programmes, yard improvements, and the deployment of new equipment, many ports have achieved great improvement on crane moves per hour. In recent years, tandem-lift, triple-lift and even quad-lifts cranes have been installed on more and more terminals. However, these new cranes have not revolutionized the industry.[3]

3.5. *UNCTAD liner shipping connectivity index 2009*

The UNCTAD Liner Shipping Connectivity Index, or the LSCI, is a famous measure used by the United Nations to assess different countries' connection to world markets and their import and export activities. Figure 6 shows the trends in LSCI indicators from 2004 to 2009.

There are 5 indicators in the LSCI: the TEU capacity of the largest container vessels, the number of liner companies per country, the TEU capacity deployed per country, the number of services per country, and the average number of vessels per country. We can see that the ecomonic crisis has caused a significant drop in the last three indicators. For the number of liner companies, we can see that there is a decreasing trend from 2005, indicating a large number of merger and acquisition activities.[3]

4. Conclusions

The container shipping industry has enjoyed unprecedented growth in the past 20 years. The economic crisis of 2007 to 2010 has caused substantial contraction in world containerized trade and, at the meantime, unveiled several problems in container shipping industry, including oversupply of tonnage, sudden drop in demand, and declining freight rates.

The recent development of container shipping industry has shown the following trends:

(1) The size of the largest container vessel and the average vessel size are both increasing. Many carriers are investing in mega-containerships with carrying capacity over 12,000 TEU. This trend causes ports to also investing in advanced equipments to cater for the mega-vessels. Furthermore, in order to achieve higher load factor for the mega-vessels, as well as to take advantage of economies of scale, many carriers are adopting more of a hub-and-spoke structure in their shipping network.

(2) Transshipment handling has become more and more significant globally. The container transshipment volume has been growing steadily to about 25% of total container handling activities. As a result, many ports are investing in port infrastructure and productivity improvement projects to attract more transshipment demand.

(3) Global container terminal operators are increasing their market share. It is often believed that global operators have higher expertise and better equipments and capital than smaller terminal operators.

(4) Liner companies are adopting more rigorous measures to reduce cost and stabilize freight rate. These measures include slow steaming, rerouting, and holding regular meeting with customers.

During the recovery after the crisis, the above trends will continue to affect the development of the container shipping industry. There are both challenges and opportunities for liner companies and terminal operators in the post-crisis era.

References

1. Sánchez, Ricardo J. The Crisis: Long Run Perspective, Signals of Over-Tonnage, and Crazy Pricing. Maritime Bulletin. 2009. Vol. 36.
2. The World Fact Book. Central Intelligence Agency Web site. [Online] June 24, 2010. [Cited: July 20, 2010.] https://www.cia.gov/library/publications/the-world-factbook/geos/xx.html.
3. The UNCTAD Secretariat. Review of Maritime Transport. Geneva: s.n. 2009.
4. Field, Alan M. Uploading a Recovery. The Journal of Commerce Magazine. Janurary 11, 2010.
5. JOC Staff. Analyst Forecasts 11 Percent Ocean Trade Growth. The Journal of Commerce Online. [Online] April 13, 2010. [Cited: July 26, 2010.] http://www.joc.com/maritime/analyst-forecasts-11-percent-ocean-trade-growth.
6. — Downturn Not Deep Enough, Says Shipping Executive. The Journal of Commerce Online. [Online] April 13, 2010. [Cited: July 26, 2010.] http://www.joc.com/maritime/Downturn-not-deep-enough-says-shipping-executive.
7. Drewry Shipping Consultants Ltd. Annual Review of Global Container Terminal Operators. s.l.: Drewry Publishing, 2009.
8. Transport and Tourism Division in UNESCAP. Regional Shipping and Port Development: Container Traffic Forecast 2007 Update. New York: United Nations, 2007.

CHAPTER 3

SCENARIO ANALYSIS FOR HONG KONG PORT DEVELOPMENT UNDER CHANGING BUSINESS ENVIRONMENT

Abraham Zhang[*][†] and George Q. Huang

*Department of Industrial and Manufacturing Systems Engineering,
The University of Hong Kong, Pokfulam Road, Hong Kong
†abraham.zhang@hkusua.hku.hk

Hong Kong port (HKP) had been the world's busiest container port during the 1990s and early 2000s. However, in recent years, its growth slowed down due to rising competition from mainland ports. This paper identifies that potential relocation of its key cargo source, processing trade enterprises, may also have fundamental impacts on HKP development. Scenario analysis is conducted to understand the relationships between business environment factors and potential relocation trends using a mixed integer programming (MIP) model. It suggests production operations are likely to move to Pan-PRD and lower-cost areas of Guangdong if business environment does not deteriorate much further. However, continual appreciation of Chinese currency RMB and further reduction of value-added tax (VAT) rebate in China will make Asian lower-cost countries more competitive. Very low oil prices will favor Inland of China. Very high oil prices will cause global manufacturing move near major markets. Processing trade relocation to western Guangdong will be slightly favorable to HKP development, while other relocation scenarios will adversely affect HKP development.

1. Introduction

Hong Kong port (HKP) had been the world's busiest container port during 1990s and early 2000s. However, dramatic changes happened in recent years. In 2005, Singapore overtook Hong Kong as the world's busiest container port. In 2007, Shanghai surpassed Hong Kong for the first time in terms

Table 1   Port statistics on container throughput (Million TEU).

Port	2003	2008	AAGR(%)
Singapore	18.4	29.9	10.2
Shanghai	11.3	28.0	19.9
Hong Kong	**20.4**	**24.5**	**3.7**
Shenzhen	10.7	21.4	15.0
Busan	10.4	13.4	5.2
Dubai	5.2	11.8	18.0
Guangzhou	2.8	11.0	31.8
Ningbo-Zhoushan	2.8	10.8	31.3
Rotterdam	7.2	10.8	8.4
Qingdao	4.2	10.4	19.9

Source: Marine Department, Hong Kong SAR (2009).

of container throughout. In terms of annual average growth rate (AAGR), Table 1 shows that HKP has had the lowest growth rate among the world's top 10 container ports from 2003 to 2008. If the trend continues, HKP will soon be overtaken by Shenzhen port, becoming the world's third busiest container port. In addition, Guangzhou port, another neighboring port, has also been on a fast ascending track. HKP is now under unprecedented challenges to keep its role as a regional shipping hub.

The slow-down on HKP development has direct impacts on the economy of the city. Ocean shipping between seaports has been the major transportation mode in global trade. The port, thus, has been a key vehicle for the operations of the trading and logistics industries in Hong Kong. These two industries have been one of the four pillar industries of the city, driving its economic development, creating employment positions and providing growth opportunities for other industries. In terms of economic contribution, trading and logistics services account for 25.8% of the GDP in 2007. The number of employees in the two sectors was 842,200, or 24.2% of total employment (Census and Statistics Department 2009).

The classical port development theory concludes that "port growth is a function of the production outcomes of firms in the port's adjacent space — or of that space to which it is linked, either in landward space or in areas linked across water or ocean" (Robinson, 1998). Indeed, HKP development has been largely determined by the growth of processing trade industries in the hinterland it serves, that is the Pearl River Delta (PRD) region adjacent to Hong Kong. This region has grown to become the "World's Factory" of many labor-intensive products. The prosperity of processing trade in the region boosted the HKP development. In 2008, over 70% of container traffic handled in Hong Kong is related to the PRD region and its adjacent areas (Hong Kong SAR Government 2009). Most of such traffic was generated by the import and export of processing trade (GPRD Business Council 2007a). Processing trade activities in the PRD region have become vital for HKP.

In recent years, an alarming phenomenon has emerged for HKP development. In 2007 and 2008, thousands of factories have ceased their operations in the PRD region due to the pressure of rising operating cost. Local governments in the PRD region also announced industry restructuring program to discourage processing trade operations in the region (GPRD Business Council 2007b). It appears to be just a matter of time that processing trade industries will relocate. If major relocation destinations are far from Hong Kong, its port development may be devastated by the loss of cargoes. Coupled with the rising competition from Shenzhen and Guangzhou ports serving the same territory, HKP faces the risk of being marginalized over long term. This new phenomenon of the relocation of processing trade out of the PRD region has not been studied thoroughly by researchers. There is also a lack of scholarly research on its implications on HKP development. This paper intends to narrow the gap by answering the following research questions:

What are the key business environment factors driving processing trade to relocate out of the PRD region, which will in turn endanger HKP development?

What are the potential relocation destinations of processing trade?

What further development in business environment may cause each relocation trend dominant?

What are the implications of each relocation trend on HKP development?

This paper surveys government and industrial reports to answer the first and second research questions. The third research question is addressed by scenario analysis through a Mixed Integer Programming (MIP) model. Sensitivity analysis is performed for key factors so as to understand their impacts on relocation destinations. Insights on potential relocation trends are drawn from a case study on a representative processing trade enterprise (PTE) in the PRD region. Based on findings from scenario analysis, the fourth research question is answered by discussing implications from potential relocation trends.

The rest of this paper is organized as follows. Section 2 reviews literature on HKP development and MIP models. Section 3 describes key business environment factors in the PRD region and potential relocation trends. Section 4 presents a MIP model. Section 5 explains experimental design. Section 6 gives modeling results, conducts scenario analysis and summarizes findings on HKP development. Section 7 concludes the paper and identifies areas for further study.

2. Literature Review

Research studies on HKP development can be categorized by their perspectives. One of the key focuses is on port competition. An early work among them was performed by Wong *et al.* (2001). A fuzzy number-based distribution model was built to predict the cargo distribution among the three main seaports in Southern China. Modeling results suggested that HKP is most attractive and its position is less likely to be overtaken. Cullinane *et al.* (2004) reviewed the recent rapid development of mainland ports and analyzed its impact on the competitiveness of HKP. In particular, competition from Shenzhen port is discussed in details. It was suggested that HKP would retain its dominant role in the region.

However, most recent studies suggested a less optimistic future for HKP. Yap *et al.* (2006) studied the port competition on hub status in East Asia. Hong Kong, Busan and Kaohsiung ports have been dominant from the 1980s. Competition intensified as Shanghai, Shenzhen and Qingdao also emerged as hub candidates in recent years. Comparative studies were conducted between traditional hub ports and emerging hub candidates. Evidence shows that HKP has become less attractive for cargoes related with China. Yap and Lam (2006) examined the competition dynamics between the major container ports in East Asia. Results suggested that Hong Kong and Pusan are beneficiaries from inter-port competition in the region from the 1970 to 2001. However, port competition in the region would intensify as the centre of gravity of cargo volume shifts to mainland China. Mainland Chinese ports have observed rapid growth on cargo throughput and they have already attracted many direct calls by shipping lines.

Several scholars made forecast on the throughput for HKP. Fung (2002) and Hui *et al.* (2004) constructed error-correction models. Their forecast results are different from those by government. Seabrooke *et al.* (2003) forecasted the cargo growth for HKP through regression analysis. It was suggested that cargo movement would continue to increase, though at a slower pace. Neighboring ports will divert cargo from HKP, but the continuous growth of the total cargo pool in Southern China would be more influential. Lam *et al.* (2004) proposed neural network models to predict cargo throughput. It was suggested neural network models are more reliable and accurate than traditional regression analysis models.

Several research studies discussed factors affecting HKP development. Loo and Hook (2002) examined the interaction among international, national and local factors. It suggested that changing policies of Hong Kong government have not been consistent with the need of a more integrated port-inland distribution system. Fung *et al.* (2003) identified that the separation of terminal handling charges (THC) from ocean freight rates since 1991 adversely affected the throughput of HKP. Cheung *et al.* (2008) used HKP for a case study on drayage services between a cargo terminal and the origin or destination of cargo. Challenges and issues in the cross-border logistics flows were highlighted. Results from an attribute-decision model suggested that relaxing policy regulations could be very beneficial. All these papers unanimously acknowledged that the manufacturing industries in the PRD region are the key cargo source for HKP. However, none of them addressed the impacts from potential relocation of manufacturing activities as a result of changing business environment.

Location decisions of global manufacturing operations often employ MIP techniques (Wilhelm *et al.*, 2005). Several researchers reviewed location decision models (Vidal and Goetschalckx, 1997; Pontrandolfo and Okogbaa, 1999; Meixell and Gargeya, 2005; Melo *et al.*, 2009). Several MIP models discussed explicitly the impacts of changes in factors.

An early MIP model was formulated by Hodder and Jucker (1985). Their model incorporated mean and variance of prices and exchange rates to address uncertainties. Modeling results suggested that it is important to address uncertainties in the international context.

Another early MIP model was developed by Cohen and Lee (1989) for a personal computer manufacturer. Its objective function is to maximize after-tax profit. It incorporated international factors including tariffs, duties and transfer pricing. Sensitivity analysis was performed to understand impacts of uncertainties in the foreign exchange rate and market demand.

Vidal and Goetschalckx (2000) analyzed impacts of uncertainties on global supply chains through a MIP model and sensitivity analysis. The paper reviewed mathematical modeling methods and suggested sensitivity analysis as the best way to analyze system variations. Uncertainties in four factors were addressed, including exchange rate fluctuation, demand change, supplier reliability and raw material shipping lead time variance. Modeling results suggested that uncertainties significantly affect optimal global supply chain configurations.

Mohamed (1999), and Mohamed and Youssef (2004) built MIP models to address impacts of exchange rates and initial capacity levels on production, distribution and investment decisions and operating profit. Modeling results were obtained for several experimental scenarios and comparative analysis suggested that both factors have significant effect.

Wilhelm *et al.* (2005) built a MIP model for a hypothetical laptop computer assembly operations in the Texas-Mexico environment under North America Free Trade Agreement (NAFTA). Impacts of business decisions, exchange rates and government investment incentives are assessed through experimental cases and sensitivity analysis.

In general, the above MIP models concluded significant impacts of various factors on global manufacturing location decisions. Sensitivity analysis is effective and beneficial for the analysis. Unfortunately, these models cannot be adapted easily for the relocation decisions of PTEs in the PRD region. None of the above MIP models was for the business environment of a developing country like China. Their relevant business models were also distinctly different from processing trade, which is dominant in the PRD region.

3. Changing Business Environment for HKP

3.1. *Changing business environment in the PRD region*

In the past five years, business environment has deteriorated rapidly for PTEs in the PRD region. Market changes have driven up business operating costs dramatically. First, Chinese currency RMB appreciated more than 20% against USD from 2005 to 2008. This appreciation has directly impacted the cost competitiveness of products made in the PRD region. Second, labor cost in the PRD region has been at sharp rise. From 2004 to 2008, the minimum wage standard was revised three times and the lowest salary for production workers increased by about 70%. In early 2008, the implementation of a new labor contract law pushed up labor cost further by an average of 23.5%, as China government mandated better social security and job security for employees. Third, oil price has been highly volatile since 2004. In July 2008, oil price hit its historical high level of about $150 per barrel, while that before 2003 was about only $30 per barrel. High oil prices led to high transportation costs, which discourages the intensive global trade activities of PTEs. Fourth, prices of many industrial raw materials have soared in correlation with that of oil price, as they are by-products of oil refinement, or incur high energy consumption during production processes. Last, utility costs have gone up too, as they are also affected by the rise of oil price (HKTDC 2008b).

Industrial policies have also changed to be in less favor of PTEs. At central government level, China reduced export value-added tax (VAT) rebates in both 2006 and 2007. It affected a very large category of labor-intensive products. "Prohibited" category has expanded for processing trade. For the production of these products, PTEs now have to make full payment of tariffs and VAT for imported raw materials and parts. "Restricted" category has also expanded. For products in the category, PTEs are subject to the customs duty deposit system. Their cash flows are affected (HKTDC 2007a). At local government level, Guangdong province started its industrial restructuring program as an attempt to upgrade its industries toward heavy, chemical and high-tech industries. Further set-up of labor-intensive and energy-intensive PTEs is discouraged in the PRD region (GPRD Business Council 2006). With the implementation of more strict environmental protection standard, sewage treatment charges more than tripled from RMB 0.25 per tonne to RMB 0.8 per tonne (HKTDC 2007a).

Among the above changes, four factors have been most influential on production costs and business profits. These four factors are RMB currency exchange rate, labor cost in the PRD region, oil price and export VAT rebate in China (HKTDC 2007a, HKTDC 2008b). As PTEs are now under tremendous pressure to relocate, HKP development is at risk due to its heavy dependence on cargoes from PTEs. If the traditional processing trade industries relocate far away from the PRD region, it is less likely that they will continue to use HKP for their import and export. On the other hand, the port and logistics facilities of Hong Kong were primarily developed for containers. They will not be able to provide the logistics support needed by the heavy and chemical industries promoted in Guangdong. Coupled with the more fierce competition from neighboring ports, HKP faces a very challenging future (GPRD Business Council 2006).

3.2. *Potential processing trade relocation trends*

Relocation involves various business considerations. Low labor costs are crucial for the production of labor-intensive products. Availability of skilled labors is also important. Exchange rate risks should be minimized. Adequate infrastructure is essential for manufacturing and logistics operations. Conducive business environment and investment incentives are also valued by global manufacturers. In addition, companies often locate within existing industrial clusters where necessary raw materials and services are readily available. Using footwear products serving the US market as an example, this research identifies five popular relocation destinations:

Guangdong (Qingyuan, Guangdong, China): The location represents lower cost areas in Guangdong province out of the PRD region (GPRD Business Council 2007b). The PRD region is in the central part of Guangdong and consists of nine counties. They are Zhaoqing, Jiangmen, Zhuhai, Foshan, Zhongshan, Guangzhou, Dongguan, Shenzhen and Huizhou. Their locations can be seen in Fig. 1. The north, west and east parts of Guangdong are less developed. Their labor costs are lower in comparison with that of the PRD region. Most of these areas are within 3 hours driving distance to the seaports in the PRD region. Relocation to these areas maintains the connectivity with existing industrial clusters in the PRD region. Many local governments offer tax and land incentives.

Pan-PRD (Ganzhou, Jiangxi, China): The location is within Pan-PRD region and 450 km away from the PRD region. The location of Jiangxi can be seen in Fig. 2. It represents lower cost areas in Pan-PRD (GPRD Business Council 2007b, HKTDC 2007b, HKTDC 2008a). The local government offer tax and land incentives.

Fig. 1   Map of Guangdong province of China.

Fig. 2   Map of Pan-PRD of China.

Inland (Bishan, Chongqing, China): The location is about 2,500 km away from the east coast of mainland China. Its labor cost is among the lowest in China. It represents other inland provinces of China (GPRD Business Council 2007b). Several large footwear plants have been established in the region to serve domestic market. Skilled labors and raw materials are available.

India (Chennai, Tamil Nadu, India): The location has easy access to a container port. It represents lower cost countries in Asia (HKTDC 2008a). The location hosts one of the largest footwear clusters in the country. Its local government offers very attractive investment incentives for export oriented manufacturing operations.

Mexico (Leon, BC, Mexico): The location is about 600 km away from the US border. It represents low cost countries which are near major international markets. A footwear cluster exists in the region. The location incurs lowest transportation costs and zero tariffs to serve the US market. It is least vulnerable to risks from high oil prices (Rubin and Tal, 2008).

PTEs' choices of relocation destinations are expected to have fundamental impacts on HKP development as they serve as the key cargo source. If PTEs relocate to Asian lower cost countries or near major markets, as represented by India and Mexico respectively in the example, HKP will be in a very unfavorable position. If PTEs relocate within China, their choices among Guangdong, Pan-PRD and Inland will cause different implications on HKP development. The routing patterns of cargo are influenced by several key factors, including port connectivity, transportation modes, transport distance, shipping cost and port competition. Generally speaking, better connectivity, shorter transport distance and lower shipping cost attract more cargo. In terms of transportation modes, barging incurs substantially lower cost than land transportation modes, and thus is preferred as long as waterways are available. HKP possesses geographical advantage for cargoes from its west via barging (Lin, 2008). If the Hong Kong-Zhuhai-Macau Bridge is completed in the next few years as planned, HKP will extend its advantage to cargoes from its west via trucking. Though the port of Gaolan in Zhuhai is nearer to cargoes from the west side, it is far less established and may not be able to develop into a competitive mainline port. However, for cargoes from its north and east, the nearest seaports are well-established Guangzhou and Shenzhen ports respectively. For both waterway and land transportation, these two ports are in more convenient locations than HKP. In addition, high THCs, high trucking cost and the low efficiency of border-crossing have been causing cargo diversion from HKP to Guangzhou and Shenzhen ports (Seabrooke *et al.*, 2003; Cullinane *et al.*, 2004).

Currently, it is unclear which relocation destination will attract most PTEs. As different relocation trends cause varying effects on HKP development, it is essential to understand which trend would become dominant and which factor plays a most influential role. To answer this research question, scenario analysis is conducted in the following three sections through MIP modeling and sensitivity analysis.

4. A MIP Model

This research builds a MIP model (Zhang and Huang, 2009) to analyze potential relocation trends. As PTEs in PRD competes mainly on cost in international markets, the model assumes that PTEs stay or move to the locations leading to lowest landed cost (LC) in markets. In general, labor-intensive manufacturing in the PRD region is not capital-intensive, and thus only major variable production costs are considered in the model.

Transportation costs are assumed to be in liner correlation with oil prices. Raw materials are assumed to be equally priced and available at all candidate locations. The PTE maintains a same net profit margin no matter where it may relocates. The model is formulated by using the parameters and variables as defined below. After presentation of the model formulation, model components are described and discussed.

Parameters

M	Markets $\{1, 2, \ldots m, \ldots M\}$
F	Facilities $\{1, 2, \ldots f, \ldots F\}$
D_m	Yearly demand for the market m supplied by the PTE
DFC_f	Base domestic freight cost per unit from facility f to port
IFC_{fm}	Base international freight cost per unit from facility f to market m
RMC_f	Raw material cost per unit of production at facility f
CRC_f	Unit capacity retaining cost at facility f
RLC_f	Unit capacity relocation cost at facility f
DLC_f	Hourly direct labor cost at facility f
Hr_f	Man hour per unit of production at facility f
OP	Oil price
OPB	Base oil price
e_f	Exchange rate to USD from the currency used for facility f
$Ctax_f$	Corporate income tax rate at facility f
VAT_f	Realized VAT rate at facility f with consideration of VAT rebate
$NPAT_f$	Net profit margin after tax at facility f
IDM_f	Days of inventory accountable to facility f
IDT_{fm}	Days of inventory in transit from facility f to market m
$Tariff_{fm}$	Tariff rate imposed by market m for the product imported from facility f
ϕ	Correlation factor between freight costs and oil price
μ	Inventory holding cost coefficient
L	A sufficiently large constant

Decision variables and convenience variables

π_f	Binary value for facility f (1 if open; 0 if not open)
q_{fm}	Yearly production quantity at facility f for market m
SP_f	Unit selling price from facility f at EXW term
$MFGC_f$	Unit manufacturing cost at facility f

$DFCC_f$ Domestic freight cost per unit from facility f to port
$IFCC_{fm}$ International freight cost per unit from facility f to market m
$CTAXC_f$ Corporate income tax per unit at facility f
$VATC_f$ VAT per unit at facility f

Objective function
   Minimize:

$$\sum_f \sum_m q_{fm} SP_f (1 + IDT_{fm}\mu)$$
$$+ \sum_f \sum_m q_{fm}(DFCC_f + IFCC_{fm} + Tariff_{fm}SP_{fm}) \tag{1}$$

Constraints

$$MFGC_f = (CRC_f + RLC_f + DLC_f Hr_f)e_f$$
$$+ RMC_f(1 + IDM_f\mu) \quad \forall f \in F \tag{2}$$

$$DFCC_f = DFC_f(1 + \phi(OP/OPB - 1))e_f \quad \forall f \in F \tag{3}$$

$$IFCC_{fm} = IFC_{fm}(1 + \phi(OP/OPB - 1)) \quad \forall f \in F \tag{4}$$

$$CTAXC_f = (SP_f - MFGC_f - VATC_f)Ctax_f \quad \forall f \in F \tag{5}$$

$$VATC_f = (SP_f - RMC_f)VAT_f \quad \forall f \in F \tag{6}$$

$$SP_f - MFGC_f - VATC_f - CTAXC_f = NPAT_f SP_f \quad \forall f \in F \tag{7}$$

$$\sum_f q_{fm} = D_m \quad \forall m \in M \tag{8}$$

$$\sum_m q_{fm} \geq \pi_f \quad \forall f \in F \tag{9}$$

$$\sum_m q_{fm} \leq \pi_f L \quad \forall f \in F \tag{10}$$

$$q_{fm} \text{ is an integer } \forall f \in F \quad \text{and} \quad m \in M \tag{11}$$

$$\pi_f \in \{0,1\} \quad \forall f \in F \tag{12}$$

$$\text{All variables } \geq 0 \tag{13}$$

The objective function (1) minimizes yearly LC in all international markets. LC components include selling prices of the manufacturer, freight costs and tariff costs. Equations (2)–(6) are convenience constraints to simplify model formulations while defining individual cost components, including manufacturing cost, domestic freight cost, international freight cost, corporate income tax and VAT tax. Equation (7) is a constraint on net profit margin after tax. Equation (8) is a demand constraint. Constraint (9) ensures that no facility is open without any production. Constraint (10) is a binary force constraint. Constraints (11)–(13) define the characteristics of variables.

5. Experimental Design

5.1. *Experimental scenarios*

The modeling work focuses on impacts of the four key factors as identified in Sec. 3.1. These four factors are RMB currency exchange rate, labor cost in the PRD region, oil price and export VAT rebate in China. Other factors are assumed to be stable as in 2008 constant price.

An experimental scenario is defined as a unique combination of parameter values to represent a certain business environment. Base scenario represents the best prediction on the future business environment for the planning horizon of 10 years. Table 2 shows parameter values at base scenario, which closely represents business conditions in the PRD region in early 2008. RMB to USD exchange rate is 0.143. Its labor cost is based on that of 2008. Export VAT rebate in China is 13%. Average oil price is projected to be at $75 per barrel. Scenarios for sensitivity analysis are also defined in Table 2.

5.2. *Experimental data*

The MIP model is applied through a case study with a labor-intensive PTE in the PRD region. The company produces low to middle-end sports

Table 2   Definition of experimental scenarios.

Factor	Base scenario	Scenarios for sensitivity analysis (absolute values or in relation to base scenario)
RMB to USD exchange rate	0.143	90%; 110%; 120%
Labor cost in the PRD region	As in 2008	80%; 120%; 140%
Oil price ($ per barrel)	75	37.5; 150; 225; 300
Export VAT rebate in China	13%	8%; 11%; 15%; 17%

Table 3   Manufacturing cost data.

	CRC_f	RLC_f	DLC_f	Hr_f	IDM_f	IDT_{fm}	RMC_f
PRD	¥6.85	¥0	¥12.77	0.8	35	28	$2.35
Guangdong	¥5.01	¥1.17	¥10.23	0.8	36	28	$2.35
Pan-PRD	¥4.81	¥1.33	¥9.81	0.8	37	28	$2.35
Inland	¥4.38	¥1.67	¥7.36	0.8	42	28	$2.35
India	$0.87	$0.50	$0.85	0.9	42	35	$2.35
Mexico	$1.19	$0.58	$2.17	1.0	35	7	$2.35

Table 4   Prevailing tax, tariff rates and freight costs.

	$Ctax_f$	VAT_f	$Tariff_{fm}$	DFC_f	IFC_{fm}
PRD	15.0%	4%	20.0%	¥0.288	$0.593
Guangdong	4.0%	4%	20.0%	¥0.552	$0.593
Pan-PRD	4.0%	4%	20.0%	¥1.000	$0.593
Inland	4.0%	4%	20.0%	¥4.000	$0.593
India	4.0%	0%	20.0%	$0.049	$0.752
Mexico	5.8%	0%	0%	$0	$0.208

shoes. Its business model and material flow represent most PTEs in the region.

Tables 3 and 4 present various input data in relation to the US market. The first rows give parameters corresponding to definitions in Sec. 4. Labor costs at all locations are valid as in 2008. All transportation modes use 40FT full containers or equivalent, which contains 4,800 units of finished products. Data for costs in India and Mexico are given in the USD. Major data sources are Hong Kong Trade Development Council (HKTDC), Investment Commission of India, ProMexico, Werner International, Shenzhen Container Trailer Association and four logistics service providers.

Demand of the US market for the manufacturer is 6,000,000 units per year. Correlation factor between freight costs and oil price (ϕ) is 40.4% in relation to the base oil price (Rubin and Tal, 2008). Inventory holding cost coefficient (μ) is set as 25% per year to account for the cost of holding inventory in transit or at manufacturing sites.

6. Results, Analysis and Findings

6.1. *Modeling results at base scenario*

As the numerical example considers only the US market, only a single manufacturing facility is suggested to be in operation at all scenarios. Values

Table 5   Landed cost comparison at base scenario.

Manufacturing location	Landed cost per unit	Difference to PRD
PRD	$7.06	—
Guangdong	$6.54	−7.46%
Pan-PRD	$6.53	−7.52%
Inland	$6.58	−6.90%
India	$6.68	−5.50%
Mexico	$6.94	−1.70%

of decision variable q_{fm} at all scenarios equal to the market demand. Table 5 shows modeling results of LC per unit of product at candidate locations at base scenario. It suggests highest LC savings of 7.52% can be achieved by relocation to Pan-PRD, followed by 7.46% for Guangdong and 6.90% for Inland. Inland has the lowest labor cost in China, but its freight cost is much higher than that of PRD and Guangdong. Relocation to India brings LC savings by 5.50%, which is less competitive than the three candidate locations in China. The advantage of lower labor cost in India is offset by its higher logistics cost. It is surprising that relocation to Mexico brings LC reduction of 1.70% though its labor cost is much higher than that of PRD. This is mainly due to its much lower logistics cost to the US and zero tariff benefit under NAFTA.

6.2. *Sensitivity analysis*

Figure 3 consists of four graphs to show the results of sensitivity analysis. Graphs 3(a) and (b) employee comparative values of factors in relation to base scenario, while graphs 3(c) and (d) use absolute values of factors. The comparison of graphs (a), (b), (c) and (d) suggests that impacts of RMB exchange rates and labor costs in the PRD region are most significant. For the current manufacturing operation in the PRD region, 10% RMB appreciation against base scenario would raise LC from $7.06 to $7.40. Labor cost increase by 20% would push up LC from $7.06 to $7.46. In comparison, impacts of oil prices and VAT rebates in China are less significant.

Impacts of factors on optimum relocation destinations could be dramatic. Graph 3(a) suggests that RMB appreciation of 10% against base scenario would cause all three locations in China to lose advantage to India. The cost advantage of manufacturing in China has become very slim after RMB appreciation of over 20% since 2005. Further RMB appreciation would quickly erode its cost competitiveness over lower cost countries in Asia.

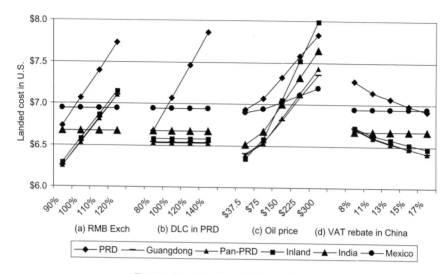

Fig. 3 Results of sensitivity analysis.

Graph 3(b) suggests that the labor cost in the PRD region has become too high for the PTE to stay competitive. Even if its labor cost decreases to 80% of the level as in base scenario, it would still not become most competitive. Continual rise of labor cost in the PRD region would further undermine its competitive position. Graph 3(c) shows that high oil prices adversely affect manufacturing locations with higher logistics costs, for example, Inland of China and India. Such effects from oil prices could dramatically alter optimum relocation destinations. If oil price stays at low level of $37.5 per barrel, Inland of China would be the optimum relocation destination due to its lowest labor cost. At base oil price of $75 per barrel, Pan-PRD is most cost competitive. At oil prices of $150 and $225 per barrel, Guangdong is most cost effective due to its lower freight costs than Pan-PRD. If oil price rises wildly to the very high level of $300 per barrel, Mexico would be the best location to serve the US market. High oil prices amplify the logistics advantage of production in Mexico. Graph 3(d) suggests that changes of VAT rebates in China between 11% and 17% do not cause the shift of optimum relocation destination. However, if VAT rebate decrease to be as low as 8%, India will become most cost competitive.

In general, manufacturing in Pan-PRD and Guangdong leads to minimal difference on LC at all scenarios. In comparison with base scenario of manufacturing in the PRD region, relocation to either of these two locations will bring about 20% labor cost savings and only incur slightly higher freight costs. Besides lowest LC, they will bring least challenges on supply chain lead time, language, culture difference, or availability of skilled labors. It is likely that Pan-PRD and Guangdong will attract a large number of relocating PTEs if business environment does not deviate far from base scenario.

6.3. *Scenario analysis and findings*

Table 6 summarizes impacts of factors on the choice of primary relocation destinations in relation to base scenario. Future scenarios only consider changes in a single factor. In most scenarios, Guangdong and Pan-PRD are the most preferred relocation destinations. Exceptions only happen in four scenarios: if RMB currency appreciates more than 10% against base scenario, Asian lower cost countries become more preferred; If oil price stays at very low level of $37.5 per barrel or below, Inland of China become most cost competitive; If oil price persists at very high level of $300 per barrel or above, Mexico would become most favorable. If VAT rebate in China

Table 6   Scenario analysis on factors and relocation trends.

Factor	Future scenario	Primary relocation destinations
RMB to USD exchange rate	Depreciate	Guangdong and Pan-PRD
	Stable	Guangdong and Pan-PRD
	Appreciate by less than 10%	Guangdong and Pan-PRD
	Appreciate by more than 10%	Asian lower cost countries
Labor cost in the PRD region (2008 constant price)	Drop slightly	Guangdong and Pan-PRD
	Stable	Guangdong and Pan-PRD
	Rise	Guangdong and Pan-PRD
Oil price per barrel (2008 constant price)	$37.5	Inland
	$75; $150; $225	Guangdong and Pan-PRD
	$300	Near major markets
VAT rebate in China	Between 11% and 17%	Guangdong and Pan-PRD
	8% or below	Asian lower cost countries

falls to 8% or below, Asian lower cost countries will gain advantage over locations in China.

It should be noted that the combined effect of multiple factors may cause the relocation dynamics more complicated. For example, both RMB currency appreciation and VAT rebate decrease cause locations in China less competitive than lower cost countries in Asia. The combined effect of these two factors would cause changes in critical levels of factors which trigger the shift of optimum locations. Taking VAT rebate as an example, its changes between 11% and 17% do not alter optimum manufacturing location if other factors stay as in base scenario. However, if RMB appreciates another 5% against base scenario, changes in VAT rebates in the same range would cause the shift of optimum manufacturing locations. Another example can be seen on the change of critical oil price causing manufacturing moving near major markets. If RMB appreciates another 10% against base scenario, oil price $170 per barrel would cause Mexico more competitive than lower cost countries in Asia.

Following the above analysis on relocation trends, Table 7 suggests their implications on HKP development. Relocation of PTEs to other parts of Guangdong province may bring mixed effect. If PTEs move to western Guangdong, their cargoes are more likely to be captured by HKP. If PTEs relocate to northern or eastern Guangdong, HKP will be less favored in comparison with Guangzhou and Shenzhen ports respectively. If PTEs move to Pan-PRD, HKP development will be in greater danger as land transportation will become the dominant mode of cargo transportation. Besides its disadvantage on cargo transported via trucking, HKP is lack of railroad linkage. It will lose opportunities to Shenzhen and Guangzhou ports when railroad becomes most economical for long distance land transportation. Such an unfavorable position will be seen for HKP if PTEs

Table 7   Implications of relocation trends on HKP development.

Relocation destination		Nearest mainline ports	Implications on HKP development
Guangdong	West	Hong Kong	Slightly favorable
	North	Guangzhou	Slightly unfavorable
	East	Shenzhen	Slightly unfavorable
Pan-PRD		Shenzhen, Guangzhou	Modestly unfavorable
Other Inland areas of China		Shanghai, Ningbo-Zhoushan, Qingdao, Tianjin	Most unfavorable
Lower cost countries in Asia		Ports in or near the location	Most unfavorable
Near major markets		Ports in or near the location	Most unfavorable

shift their operations to provinces which are far away from Hong Kong, including west part of Pan-PRD and other inland provinces. If PTEs move to lower cost countries in Asia, HKP will lose geographical proximity to cargo source. At most, it will intercept some transshipment cargo from nearby lower cost countries in Asia. Relocation to near major markets would happen when oil price stays at very high levels, which will greatly discourage global trade. HKP development, along with other ports in Asia, would become very pessimistic in this scenario.

7. Conclusions and Future Work

This research studies impacts of business environment factors in the PRD region on the port development of Hong Kong. Processing trade activities in the PRD region have been the primary cargo source for HKP. Their rapid expansion in the past two decades has boosted the role of Hong Kong as a regional shipping hub. However, in recent years, business environment for processing trade has deteriorated dramatically. Chinese currency RMB appreciated over 20% since 2005. Minimum wage limits in PRD had moved up by about 70% since 2004. The introduction of a new labor contract law in China at the start of 2008 pushed up labor cost further. Oil price has fluctuated wildly at various high levels since 2004, which leads to high transportation costs in the global trade. China central government reduced export VAT rebates for a very large category of processing trade products in both 2006 and 2007. Local Guangdong government also started industrial restructuring program to discourage processing trade activities in the PRD region. All these factors are pushing processing trade activities to relocate. If relocation destinations are far from Hong Kong, its port would be marginalized over long term.

This paper contributes to the study on HKP development. First, it points out that potential relocation of processing trade activities out of the PRD region may have fundamental impacts on HKP development. Relevant key factors and potential relocation trends are also identified. Second, a MIP model is constructed for scenario analysis. The relationships between factors and potential relocation trends are established through sensitivity analysis. PTEs are likely to move to Pan-PRD and other parts of Guangdong if business environment does not deteriorate much further from that of early 2008. Lower cost countries in Asia may become more competitive if RMB appreciates another 10% from base exchange rate of 0.143, or VAT rebate in China falls to as low as 8% or below. If oil price stays at very low level of

$37.5 per barrel or below, Inland of China may become most cost effective. If oil price persists at very high level of $300 or above, locations near major markets will become most preferred. Last, implications of potential relocation trends on HKP development are discussed. Relocation of PTEs to western Guangdong will be slightly favorable. Relocation to northern and eastern Guangdong will be slightly unfavorable. Relocation to Pan-PRD will be modestly unfavorable. Most unfavorable trends are the relocations of PTEs to other inland provinces of China, lower cost countries in Asia and near major markets.

The research is limited in several areas. First, the modeling work uses a footwear PTE for case study. Modeling results are expected to be valid in general as footwear is a typical labor-intensive processing trade product in the PRD region. However, due to differences on cost structures and freight-intensiveness, other products may have different critical levels in factors to trigger various relocation trends. Second, the MIP model focuses on the effect of four key factors. A real life manufacturing relocation decision usually involves more factors. Third, potential relocation trends are subject to uncertainties. It is not impossible that unexpected factors would alter the potential relocation trends. Last, implications on the HKP development are based on the current situation of the port and neighboring mainland ports. The changing port competition dynamics in the region may lead to different prospects.

Future work on the subject is to derive policy responses for the HKP development. This will be of practical benefits to the port and the economy of Hong Kong. As the logistics industry of Hong Kong is twinned together with the trading industry, further study on trading industry of Hong Kong will be required. Port competition from mainland ports has diverted a large portion of cargo from the HKP. Future work will include a detailed comparative study on the advantages and disadvantages of ports in the territory.

References

1. Census and Statistics Department (2009). *The Situation of the Four Key Industries in the Hong Kong Economy in 2007. Hong Kong Monthly Digest of Statistics.* Hong Kong: Census and Statistics Department, The Government of Hong Kong SAR, FB0-FB16.
2. Cheung, R., Shi, N., Powell, W. and Simao, H. (2008). An attribute-decision model for cross-border drayage problem, *Transportation Research. Part E, Logistics & Transportation Review*, 44(2), pp. 217–234.

3. Cohen, M.A. and Lee, H.L. (1989). Resource deployment analysis of global manufacturing and distribution networks, *Journal of Manufacturing and Operations Management*, 2, pp. 81–104.
4. Cullinane, K., Fei, W.T. and Cullinane, S. (2004). Container terminal development in Mainland China and its impact on the competitiveness of the port of Hong Kong, *Transport Reviews*, 24(1), pp. 33–56.
5. Fung, M.K. (2002). Forecasting Hong Kong's container throughput: an error-correction model, *Journal of Forecasting*, 21(1), pp. 69–80.
6. Fung, M.K., Cheng, L.K. and Qiu, L.D. (2003). The impact of terminal handling charges on overall shipping charges: an empirical study, *Transportation Research Part A — Policy and Practice*, 37(8), pp. 703–716.
7. GPRD Business Council (2006). Report on Guangdong's industrial restructuring — opportunities and challenges for Hong Kong. Hong Kong: The Greater Pearl River Delta Business Council.
8. GPRD Business Council (2007a). *The Greater Pearl River Delta Business Council 2006/2007 Annual Report*. Hong Kong: The Greater Pearl River Delta Business Council.
9. GPRD Business Council (2007b). *Implications of Mainland Processing Trade Policy on Hong Kong*. Hong Kong: The Greater Pearl River Delta Business Council.
10. HKTDC (2007a). Cost escalation and trends for export price increase — A look at the rising production costs in the PRD. Hong Kong: Hong Kong Trade Development Council.
11. HKTDC (2007b). Relocating processing trade from PRD — An assessment of alternative destinations on the mainland. Hong Kong: Hong Kong Trade Development Council.
12. HKTDC (2008a). Latest development and strategies of Hong Kong companies' processing trade business in the Pearl River Delta. Hong Kong: Hong Kong Trade Development Council.
13. HKTDC (2008b). Upward pressure on export prices mounting but competitiveness maintained. Hong Kong: Hong Kong Trade Development Council.
14. Hodder, J.E. and Jucker, J.V. (1985). International plant location under price and exchange rate uncertainty. *Engineering Costs and Production Economics*, 9(1–3), pp. 225–229.
15. Hong Kong SAR Government (2009). Transport. *Hong Kong Yearbook 2008*. Hong Kong: Hong Kong SAR Government.
16. Hui, E.C.M., Seabrooke, W. and Wong, G.K.C. (2004). Forecasting cargo throughput for the port of Hong Kong: error correction model approach, *Journal of Urban Planning and Development-ASCE*, 130(4), pp. 195–203.
17. Lam, W.H.K., Ng, P.L.P., Seabrooke, W. and Hui, E.C.M. (2004). Forecasts and reliability analysis of port cargo throughput in Hong Kong, *Journal of Urban Planning and Development*, 130(3), pp. 133–144.
18. Lin, W. (2008). Towards the upliftment of the Hong Kong port. Shippers Today. Hong Kong.

19. Loo, B.P.Y. and Hook, B. (2002). Interplay of international, national and local factors in shaping container port development: a case study of Hong Kong, *Transport Reviews*, 22(2), pp. 219–245.

20. Marine Department (2009). *Ranking of Container Ports of the World*. Hong Kong: The Government of the Hong Kong SAR.

21. Meixell, M.J. and Gargeya, V.B. (2005). Global supply chain design: a literature review and critique, *Transportation Research Part E*, 41, pp. 531–550.

22. Melo, M.T., Nickel, S. and Saldanha-da-Gama, F. (2009). Facility location and supply chain management — A review, *European Journal of Operational Research*, 196(2), pp. 401–412.

23. Mohamed, Z.M. (1999). An integrated production-distribution model for a multinational company operating under varying exchange rates, *International Journal of Production Economics*, 58(1), pp. 81–92.

24. Mohamed, Z.M. and Youssef, M.A. (2004). A production, distribution and investment model for a multinational company, *Journal of Manufacturing Technology Management*, 15(6), pp. 495–510.

25. Pontrandolfo, P. and Okogbaa, O.G. (1999). Global manufacturing: a review and a framework for planning in a global corporation, *International Journal of Production Research*, 37(1), pp. 1–19.

26. Robinson, R. (1998). Asian hub/feeder nets: the dynamics of restructuring, *Maritime Policy and Management*, 25, pp. 21–40.

27. Rubin, J. and Tal, B. (2008). *Will Soaring Transport Costs Reverse Globalization?* Toronto: CIBC World Markets Inc.

28. Seabrooke, W., Hui, E.C.M., Lam, W.H.K. and Wong, G.K.C. (2003). Forecasting cargo growth and regional role of the port of Hong Kong. *Cities*, 20(1), pp. 51–64.

29. Vidal, C.J. and Goetschalckx, M. (1997). Strategic production-distribution models: a critical review with emphasis on global supply chain models, *European Journal of Operational Research*, 98, pp. 1–18.

30. Vidal, C.J. and Goetschalckx, M. (2000). Modeling the effect of uncertainties on global logistics systems, *Journal of Business Logistics*, 21(1), pp. 95–120.

31. Wilhelm, W., Liang, D., Rao, B., Warrier, D., Zhu, X. and Bulusu, S. (2005). Design of international assembly systems and their supply chains under NAFTA. *Transportation Research Part E*, 41, pp. 467–493.

32. Wong, W.G., Han, B.M., Ferreira, L. and Zhu, X.N. (2001). Factors influencing container transport: a fuzzy number-based distribution model approach, *Transportation Planning and Technology*, 24(3), pp. 171–183.

33. Yap, W.Y. and Lam, J.S.L. (2006). Competition dynamics between container ports in East Asia, *Transportation Research Part A — Policy and Practice*, 40(1), pp. 35–51.

34. Yap, W.Y., Lam, J.S.L. and Notteboom, T. (2006). Developments in container port competition in East Asia, *Transport Reviews*, 26(2), pp. 167–188.

35. Zhang, A. and Huang, G.Q. (2009). A Mixed Integer Programming Model to Evaluate the Impact of Business Factors on Global Manufacturing Relocation Decisions. Proceedings of 39th International Conference on Computers & Industrial Engineering. Troyes, France, pp. 390–395.

CHAPTER 4

MODELS FOR PORT COMPETITIVE ANALYSIS IN THE ASIA-PACIFIC REGION

Chew Ek Peng*, Lee Loo Hay, Jiang Jianlin and Gan Chee Chun

Department of Industrial & Systems Engineering
National University of Singapore, Singapore, 119260
*isecep@nus.edu.sg

The trend of increasing globalization of the world's economies and the relentless development of international trade links has resulted in a great boom in many transportation industries, notwithstanding the recent economic downturn. Sea container traffic accounts for the majority of international trade due to its cost-effectiveness and as such maritime transportation is vital for the economic health of many nations. The container transportation industry has therefore experienced great growth, resulting in intense competition among international container ports. In such a competitive environment, port benchmarking has become particularly important for individual ports as it can help them recognize potential threats and opportunities as well as reveal their strengths and weaknesses in the face of the ever-changing competitive landscape.

This paper focuses on port benchmarking from the following three perspectives: port efficiency, port connectivity, and the impact of various factors on individual ports. First we present a brief review on the study of port benchmarking in recent years. Next, we examine the aforementioned three perspectives by presenting a model and a case study for each model. Efficiency is an important factor in port competitiveness, and by using Data Envelopment Analysis methodology many port attributes can be evaluated in terms of their impact on port performance. Port connectivity indicates how well ports connect to others in the actual maritime transportation network and is vital for the competitiveness of transshipment services at a port. However, to date this concept has not yet been clearly defined. A port connectivity analysis framework is proposed in this paper and under this framework this concept can be characterized in terms of the impact of transshipment operations at a port on the overall network. Lastly, many factors can influence a port's performance and competitiveness and a given factor can play different roles for different ports. The complex interplay of factors in port systems can make it difficult to study the impact of individual factors on the port. A network flow model is presented to analyze the impact of different factors on different ports. As Asia-Pacific is the pre-eminent region in terms of world container through-put, this paper focuses on the benchmarking analysis of major ports in the Asia-Pacific region in order to provide some insights into the performance of these ports.

1. Introduction

In recent years, the maritime transportation industry has seen great growth in many areas due to the globalization of trade, growth in sea transport, development of business logistics and specialization in production (Cuadrado *et al.*, 2004). As the major gateways for international trade, maritime ports are vital components of the maritime transportation network. Accordingly, their importance as economic drivers in the region has also led many countries to invest large sums of money in order to lower operational costs or improve service quality. More and more new ports have also been developed to tap into the global maritime traffic. As a result, the competition among maritime ports has also been rapidly increasing. This is of great concern for individual ports, as their competitive ranking can have a very large effect on their future well-being in such a fierce competitive environment. A decline in traffic can potentially lead to irreparable long-term repercussions. If liner companies shift services away from a port due to poor connections, high costs or poor handling quality, the loss of connections at the port may incite additional liner companies to shift away as well. This can lead to a slippery slope where ports can find it very difficult to recover their competitive advantage once it is lost.

In such an environment, it becomes vitally important for ports to conduct benchmarking studies to constantly evaluate their position within the competitive landscape. According to Bemowski (1991), benchmarking is "the measurement of a company's performance in comparison to the best, determining how those companies achieve superior performance and using that information as the basis to decide on and implement objectives and strategies". By examining the strengths and weaknesses of themselves and their competitors, ports can quickly detect any facet of operations at which they are lagging behind. This enables port managers to quickly make decisions and put in place measures to arrest the slide before too much damage is done. In this chapter, port benchmarking is used to analyze port competitiveness from three different perspectives: port efficiency, port connectivity and the impact of different factors on individual ports.

2. Literature Review

Numerous studies have been done on port competitiveness, some of which have involved benchmarking in the port sector, though not necessarily looking at only the container terminals. Various ways of evaluating or comparing ports and even container terminals have been applied. For instance, Porter's diamond has been used to identify the factors that determine the competitiveness of the ports (Rugman, 1993) and Strategic Positioning Analysis (SPA) which focuses on the market share and growth within a time period has also been proposed and applied (Huybrechts, 2002). Kleywegt *et al.*, (2002) analyzed the competition between the ports of Singapore and Malaysia. The paper compares various attributes of the two ports, considers the economic factors behind the transfer of Maersk Sealand and provokes some thinking on strategies the ports should adopt in the future. Using the benchmarking analysis, Pardali and Michalopoulos (2008) give a score for each attribute considered based on how good the ports are in that attribute and identify the best port as being the one which has the highest number of maximum scores. The port competitiveness degree (PCD) is then considered to further test the result. The ranking of all other ports is obtained through comparison to the best port in the model. According to the authors, this methodology has the advantage of taking into consideration an unlimited number of factors.

The factors which can influence a port's efficiency are complex and the correlation between those different factors often cannot be determined exactly. Nevertheless, the Data Envelopment Analysis (DEA) methodology is an ideal tool to deal with such problems due to its attributes. In fact, in the last decade, there has been a continuous and steady increase in the use of DEA for port benchmarking in terms of the efficiency of individual ports. The reason for such an increase is that the methodological and computational benefits of DEA make it rather suitable for efficiency measurement in the complicated port environment. The work of Roll and Hayuth, (1993) was the first attempt to use DEA to measure port efficiency and thus it can be regarded as a milestone of DEA application to seaport efficiency measurement, although it was purely theoretical and did not use any actual data. Poitras *et al.*, (1996) analyzed the relative efficiency of 23 Australian and international ports with both the CCR model and the BCC model. Martinez-Budria *et al.*, (1999) used DEA to measure the efficiency of 26 Spanish ports. Tongzon, (2001) applied DEA to measure the relative efficiency of some Australian and other international ports. Both constant and variable returns to scale models (CCR and ADD) were chosen to analyze the selected ports and the author concluded that the CCR model would result in more inefficient ports than the ADD model, and that the number of the selected ports should not be too small. Valentine and Gray, (2001) adopted DEA to analyze the relationship between port ownership structure and efficiency. Barros, (2003) used DEA to analyse 11 ports of Portugal using panel data between 1990 and 2000. The author concluded that the most important factors of a port are its dimensions and location while capital intensity and private ownership offered little significant impact to port efficiency. Park and De, (2004) presented a four-stage DEA model to measure port efficiency, in which the overall efficiency is divided into four stages, namely productivity, profitability, marketability and overall efficiency. Cullinane et al. (2006) compared two methods for port efficiency measurement: DEA and SFA. This research finds that the two methods would yield similar efficiency scores in terms of the ranking of the ports. Rather recently, Sharma and Yu, (2009) combined data mining and DEA to present a diagnostic tool to measure the efficient terminals and prescribe a step-wise projection to reach the frontier according to their maximum capacity and similar input properties.

Port connectivity has also been examined from a few aspects during the last several years. Hoffmann, (2005) combined 10 indicators of maritime transportation, including fleet assignment, liner services, and vessel and fleet sizes and so on, to generate an overall liner shipping connectivity indicator for each country. McCalla et al., (2005) measured the connectivity for Caribbean shipping networks. Notteboom, (2006) discussed the time factor in liner services which reflects one port's connectivity. Marquez-Ramos et al., (2006) used principle component analysis (PCA) methodology to build three complex connectivity component variables and analyze the determinants of maritime transport costs of Spanish exports and their effect on international trade flows. Wilmsmeier et al., (2006) investigated maritime trade among 16 Latin-American countries and their findings revealed that inter-port connectivity has a significant impact on international maritime transport costs. Wilmsmeier and Hoffmann, (2008) analyzed the impacts of liner shipping connectivity on intra-Caribbean freight rates and the relationships between the structure of liner services, port infrastructure and liner shipping freight rates. Tang et al., (2008) and Low et al., (2009) proposed a direct measure of port connectivity based on the number of origin and destination pairs served by individual ports in real transportation networks.

Port selection criteria of carriers and shippers have been studied for several decades and the impacts of various factors on ports in different regions have been investigated in depth. Slack, (1985) analyzed the containerized traffic between the North American Mid-West and Western Europe, from which the author concluded that the price and service are the most important factors. Murphy *et al.*, (1992) developed a framework to analyze port selection factors for different maritime players. Tiwari *et al.*, (2003) used a discrete choice model to simulate port choice behavior from which they find that the distance to destination, the distance from origin, port congestion, and shipping line's fleet size play an important role in port selection. Malchow and Kanafani, (2004) present an alternative form of the discrete choice model to analyze the distribution of maritime shipments among US ports and found that the most significant factor of a port is its location. Recently, Tongzon and Heng, (2005) summarized eight key determinants of port competitiveness from the existing literature, namely port operation efficiency level, cargo handling charges, reliability, port selection preferences of carriers and shippers, the depth of the navigation channel, adaptability to the changing market environment, landside accessibility, and product differentiation. Analytic Hierarchy Process (AHP) is a multi-objective, multi-criteria theory of measurement created by Saaty (1977) and it has already been employed to determine the predominant factors in port selection decisions. Lirn *et al.*, (2004) use AHP to analyze and reveal important service factors for transshipment port selection by global carriers. Song and Yeo, (2004) employ AHP to identify competitiveness of container ports in China and provide managerial and strategic implications. By using AHP, Ugboma *et al.*, (2006) determine the factors that carriers and shippers consider important. Their findings show that the carriers and shippers place high emphasis on efficiency, frequency of ship visits and adequate infrastructure.

3. Port Benchmarking Models

Many effective port benchmarking methodologies exist to evaluate different aspects of port operations. This chapter focuses on evaluating ports from the perspective of port efficiency, port connectivity and the impact of factors on port performance. Accordingly, a brief overview of 3 separate models is presented. Port efficiency will be evaluated using a Data Envelopment Analysis (DEA) model. A framework is then proposed for the analysis of port connectivity from a network perspective. Lastly, a network flow model based on the work of Lee *et al.* (2006) is used to evaluate the impacts of various factors on different ports. To illustrate these methodologies, a case study is performed on various ports in the Asia Pacific region.

3.1. *Port efficiency*

Maritime transportation is vital to the health of a country's economy, as more than 85% of international trade is transported through seaports (Liu, 2008). Coelli, (1996) indicates that there are mainly two methods used to measure efficiency: the stochastic frontiers approach (SFA) and the Data Envelopment Analysis (DEA) approach. SFA utilizes regression analysis to determine the inefficiency values and the factors which have an impact on efficiency. This unfortunately requires the functional form of the regression equation to be known in advance, which can result in biases introduced by subjective recognition.

The DEA methodology measures the relative efficiency of a set of homogeneous entities called Decision Making Units (DMUs). DEA can account for the impact of numerous factors and does not require prior assumptions such as the relationship between inputs and outputs, or the weightings for each factor. As DEA is not limited to the operating types of the objects being evaluated, it can be used in a wide variety of fields. The DEA methodology is based on generating a frontier which is composed of the best DMUs, meaning those that have the greatest output/input ratio. By measuring the distance of each DMU from the efficient frontier, an efficiency value between zero and one can be assigned. DEA can also identify a set of weights that maximizes the efficiency score of a DMU while keeping the efficiency scores of other DMUs less than or equal to 1 using the aforementioned weights. However, care must be taken in the selection of the appropriate inputs and outputs as the DEA methodology does not assign weights based on the actual importance of the attributes to the final performance. DMUs can be classified as being efficient (weakly or strongly) if their efficiency score is 1. Otherwise, they are classified as being inefficient.

The last decade has seen a steady increase in the DEA's popularity as a measure of port efficiency. More details about the application of DEA methodology in port economics can be referred to in the work of Panayides *et al.*, (2009). In this section we will focus on the performance of the container terminals of the major Asian ports and DEA will be used to benchmark the major ports in the Asia Pacific region. Furthermore, the key attributes that contribute to the success of the ports will be identified.

The model is constructed using $n\,DMU_o (o = 1, \ldots n)$ with m inputs and s outputs. $x_o \in R_+^m$ and $y_o \in R_+^s$ represent the input and output vectors of DMU_o respectively, while $X = (x_o) \in R_+^{m \times n}$ and $Y = (y_o) \in R_+^{s \times n}$ denote the input and output matrices. Based on X and Y, the DEA model can be given by the following linear programming problems

Min $\quad \theta$

$$
\begin{aligned}
s.t. \quad & X\lambda \le \theta x_o \\
& Y\lambda \ge y_o \\
& (e^T \lambda = 1) \\
& \lambda \ge 0,
\end{aligned}
\tag{1}
$$

Max $\quad e^T s^- + e^T s^+$

$$
\begin{aligned}
s.t. \quad & s^- = \theta^* x_o - X\lambda \\
& s^+ = Y\lambda - y_o \\
& (e^T \lambda = 1) \\
& s^- \ge 0, \quad s^+ \ge 0, \quad \lambda \ge 0.
\end{aligned}
\tag{2}
$$

where θ and λ are the decision variables among which the former is a scalar while the other is a column vector $(\lambda_1, \ldots, \lambda_n)^T$. e is a column vector in R^n with all elements equal to 1. The optimal solution of (1), denoted by θ^*, is employed as the technical efficient score of DMU_o, which is used in (2).

If DMU_o satisfies (i) $\theta^* = 1$; (ii) $s^- \ge 0$ and $s^+ \ge 0$, then it is called fully efficient; otherwise, if DMU_o only satisfies (i) then it is called weakly efficient. In (1) and (2) $(e^T \lambda = 1)$ means this constraint can either be included or excluded, which corresponds to different DEA models. If it is excluded then (1) is called the CCR model which stipulates variable returns to scale and the units satisfying (i) and (ii) are BCC-efficient. Let θ_{CCR}^* and θ_{BCC}^* be the CCR score and BCC score of a DMU_o, respectively. Then the scale efficiency is derived by

$$
SE = \frac{\theta_{CCR}^*}{\theta_{BCC}^*},
\tag{3}
$$

SE is not greater than one since θ_{CCR}^* is always not greater than θ_{BCC}^*. $SE = 1$ indicates scale efficiency and $SE < 1$ indicates scale inefficiency.

Both the CCR and BCC models given by (1) are input-oriented which means they minimize inputs while satisfying the given output levels. There is another type of CCR and BCC model, called output-oriented, that attempts to maximize outputs while using no more than the given input levels. Besides the CCR and BCC models, there are several other classical DEA models for efficiency measurement, such as the additive model, slacks-based measure model, Russell measure model and so on. As we are only focused on using DEA models in measuring the efficiency of major ports in the Asia-Pacific region, we will not delve into the details of these other DEA models.

3.1.1. *Case study*

The model detailed in the previous section was applied to 9 major ports in Asia: Singapore, Shanghai, Hong Kong, Shenzhen, Busan, Kaohsiung, Qingdao, Ningbo, and Tanjung Pelepas. The listed ports were chosen as they are some of the largest and fastest growing ports in the world, and thus benchmarking on them should be representative and meaningful. The port data used was obtained from CI-Online, 2008.

Several scenarios were constructed in order to analyze different aspects of port operations. All the scenarios have the same input variables (Area of Container Terminals, Quay Length, No. of Quay Cranes, Storage Capacity). However, their output variables are different combinations of Derived Revenue, Port Calls/week and throughput, as shown in Table 1.

The input-oriented CCR DEA model is chosen as our benchmarking analysis tool and the efficiency scores of the nine major ports in different scenarios are shown in Table 2:

By comparing the results of Scenario 1 and Scenario 2, we can observe the effect of excluding *Derived Revenue* as an input on the resulting efficiency scores. The ports of Shenzhen and Tanjung Pelepas suffer a decrease in their efficiency score, which shows that they are more efficient at bringing in revenue with more throughput or higher cost. In effect, they are able to have higher throughput or higher cost or both with the same input or they are able to have the same throughput and cost with less input. As for Busan and Qingdao, the exclusion of *Derived Revenue* does not have an impact on their inefficient scores. This probably indicates that

Table 1   The attributes used in our study.

		Scenario 1	Scenario 2	Scenario 3	Scenario 4
Output Variables	Derived Revenue	✓		✓	
	Port Calls/week	✓	✓		
	Throughput				✓

their efficiency score in *Derived Revenue* is exactly the same as their overall score. The alternative is that the attribute does not have any impact on their overall score, which is highly unlikely.

Comparing the results of Scenario 1 and Scenario 3 provides insight on the contributions of the attribute *Port Calls/week*. It is found that the efficiency scores of 5 ports (Busan, Kaohsiung, Qingdao, Ningbo and Tanjung Pelepas) drops because of the exclusion of the attribute *Port Calls/week*, which shows that they are doing a better job in bringing in port calls, i.e., they are able to have higher port calls with the same input or they are able to have the same port calls with less input. The efficiency score of the port of Kaohsiung experiences the biggest drop, 86.57%, which means Kaohsiung is doing the best in bringing in port calls. The next are Pusan, Ningbo, Qingdao and Tanjung Pelepas.

The attribute of *Throughput* is often chosen as an output in many DEA studies, and is sometimes the only output considered. Thus it is interesting to compare the difference between *Derived Revenue* and *Throughput*, which is carried out by the comparison of Scenario 3 (*Derived Revenue* is the product of *Throughput* and *Normalized handling cost*) and Scenario 4. Ports of Singapore, Shanghai, Hong Kong and Shenzhen are efficient in both scenarios, which shows that they excel at bringing in throughput. Inefficient ports in Scenario 3 remain inefficient in Scenario 4, but they all experience an increase in their efficiency scores, which shows they are doing worse in terms of handling cost when compared with the efficient ports. Kaohsiung has the greatest increase, followed by Busan, Tanjung Pelepas, Ningbo and Qingdao.

According to Table 2, it is clear that Busan, Kaohsiung, Qingdao and Ningbo are more efficient in bringing in port calls while Shenzhen and Tanjung Pelepas are better in bringing in revenue. The three ports

Table 2   The efficiency score of different scenarios.

Port	Scenario 1 score	Scenario 2 score	Scenario 3 score	Scenario 4 score
Singapore	1	1	1	1
Shanghai	1	1	1	1
Hong Kong	1	1	1	1
Shenzhen	1	0.532664	1	1
Busan	0.917816	0.917816	0.352235	0.644344
Kaohsiung	1	1	0.134321	0.84481
Qingdao	0.638416	0.638414	0.545662	0.596209
Ningbo	1	1	0.681831	0.959301
Tanjung Pelepas	0.398707	0.373839	0.395433	0.666486

Table 3 The efficiency score for turnaround time.

Port	Scenario 1 score
Singapore	1
Shanghai	0.638757
Hong Kong	1
Busan	0.51116
Tanjung Pelepas	0.35266

Singapore, Shanghai and Hong Kong remain strongly efficient in all the 4 scenarios. This suggests that the 3 ports are likely to be the most efficient ones among all the ports studied.

A similar analysis can also be performed using *Turnaround Time* as the output, with the inputs normalized using throughput handled in order to remove the bias against large ports. Due to unavailability of data, the analysis could only be performed on 5 ports, namely Singapore, Shanghai, Hong Kong, Busan and Tanjung Pelepas.

The results for Turnaround Time (Table 3) show that Singapore and Hong Kong are the most efficient, while Shanghai is in 3rd place. This reinforces the earlier findings that Singapore, Hong Kong and Shanghai are the most efficient out of those ports included in the study.

3.2. *Port connectivity*

Transshipment operations have rapidly grown in importance in the past few decades. According to a report from Drewry Shipping Consultants, transshipment volume as a proportion of total volume handled has grown from 18% in 1990 to 28% in 2008. As such, it is no surprise that any thorough study of port competitiveness should include their transshipment capability. A major factor influencing transshipment capability is the concept of port connectivity. Intuitively, port connectivity indicates how well one port connects to others in a maritime transportation network and its ease of accessibility by liner services. A port with a high level of connectivity is likely to have a great advantage for transshipment operations, so much so that connectivity could be used as a direct proxy for a port's competitiveness in terms of transshipment. In general, the higher connectivity level a port has, the more attractive it will be to liner companies in the sense of facilitating the transportation of cargo and reducing transportation cost and time, leading to the port being more competitive than its peers.

However, it is not immediately evident how to determine a port's connectivity as it is a concept that can have different interpretations and meanings in different cases. To date, this concept of port connectivity has not been well defined despite several papers on the topic. In this section, we propose a connectivity analysis framework from a network perspective that can be used to benchmark the competitiveness of various ports in the network in terms of the impact of their transshipment operations.

At first glance, a possible way to measure the connectivity of a port is to simply count the number of direct connections it has to other ports in the network, i.e., the number of incoming and outgoing shipping services to/from the port in question. Despite providing a reasonable estimate, this simplistic method can leave out some important details. Three such important factors are discussed as follows for an origin port A and destination port B:

Responsiveness: The average waiting time that a supplier has to wait for a service to ship his goods from port A to port B. This factor is related to the frequency of shipping between port A and port B. The lower the frequency, the longer the expected waiting time for the supplier. Vice versa, the higher the frequency, the more trips from A to B is made per time period and thus the shorter the expected waiting time.

Capacity: Even if there are frequent services between port A and port B, it could be that the ships serving port A and port B have low capacities, which limit the amount of cargo that the supplier can ship at any time. On the other hand, if services between port A and port B have low frequencies but high capacities, a large amount of cargo can be shipped in a single shipment even though the average waiting time might be long. Therefore, capacity should also be considered as a factor in port connectivity.

Network structure: Besides direct links, there may also be services connecting ports A and B that comprise of multiple stops. In this case, the structure of the transportation network, such as the existence of bottlenecks or the ease of accessibility to large hubs, can also play a very large part in the viability of transshipment services and hence affect a port's connectivity.

In order to account for these factors, a new definition of connectivity is proposed as follows. Considering that transshipment has become a major port service, the connectivity of a port can be defined as the impact on the transportation network's performance when transshipment services are not available at this port. Intuitively, such impact is proportional to the connectivity of the corresponding port. This methodology also accounts for any network effects that may influence services between ports that are not directly connected. If a port has a high degree of connectivity, then many carriers will come to this port to transship their cargoes. If said port's transshipment capabilities are now disabled, then many of these carriers will have no choice but to select other ports at which they can transship their cargoes, which will likely result in greater transportation cost or longer shipping time. Thus, ports with a high degree of connectivity will result in a greater impact on the transportation network's performance than those with low connectivity when transshipment services are not available.

Such an impact on the network can take many forms. For example, the model can measure the impact on the transportation flow capacity of the system, the impact on the transportation time of cargoes and so on. It is necessary to examine the results from different perspectives in order to obtain a thorough and comprehensive understanding of this concept and provide meaningful benchmarking results.

3.2.1. *Case study*

This subsection focuses on the effects of transshipment on container throughput and shipping time of major ports in the Asia Pacific region. By comparing various scenarios in which transshipment is disabled for certain ports against a base case, we can rank the ports analyzed according to the network benefits provided by transshipment services. The network was constructed using data from the top 10 largest liner companies in 2008 according to CI-Online.

When considering the throughput model, the disabling of transshipment services at Singapore, Shanghai, Dalian and Qingdao has the greatest impact in descending order (shown in Table 4). Singapore is the largest by a wide margin as it serves as the primary transshipment hub for all traffic to Oceania and between Asia and Africa/Europe. Shanghai is close behind due to its share of transshipment traffic to the USA, while Dalian and Qingdao serve as gateway ports to the North China region. The disabling of transshipment services at Qingdao and Dalian has a significant impact on the flow of traffic to Tianjin and the Beijing area. In comparison, Hong Kong and Shenzhen have a relatively small impact on the network, especially considering that they have a very large number of linking services.

The waiting time model provides some different insights as seen in Table 5. Disabling transshipment services at Singapore still has the greatest

Table 4   Impact on throughput.

Port	Weekly TEUs per OD	Change in throughput	% change
Base case	235697.6		
Singapore	230484.3	-5213.32	-2.21
Shanghai	232170.3	-3527.29	-1.50
Dalian	232577.5	-3120.17	-1.32
Qingdao	232759.1	-2938.56	-1.25
Yokohama	233312.9	-2384.69	-1.01
Ningbo	233785.5	-1912.16	-0.81
Busan	233826.1	-1871.53	-0.79
Kaohsiung	234298.4	-1399.24	-0.59
Port Klang	234361.1	-1336.51	-0.57
Tanjung Pelepas	234838.7	-858.878	-0.36
Hong Kong	234739.9	-957.743	-0.41
Tianjin	234746.1	-951.491	-0.40
Shenzhen	234770.8	-926.856	-0.39
Guangzhou	234946.6	-751.026	-0.32
Yingkou	235334.8	-362.835	-0.15
Laem Chabang	235463.6	-233.981	-0.10

Table 5   Impact on waiting time.

Port	Waiting time per OD (days)	Change in waiting time	% change
Base case	6.924952		
Singapore	7.451377	0.526424	7.60
Busan	7.329748	0.404796	5.85
Yokohama	7.22962	0.304668	4.40
Qingdao	7.107357	0.182405	2.63
HongKong	7.04775	0.122798	1.77
Shanghai	7.029669	0.104717	1.51
Shenzhen	6.996822	0.071869	1.04
PortKlang	6.991358	0.066406	0.96
Kaohsiung	6.983623	0.05867	0.85
LaemChabang	6.980875	0.055923	0.81
Dalian	6.958633	0.033681	0.49
Ningbo	6.957301	0.032349	0.47
TanjungPelepas	6.945159	0.020207	0.29
Guangzhou	6.929884	0.004932	0.07
Yingkou	6.925413	0.00046	0.01
Tianjin	6.924952	0	0.00

impact by a wide margin due to the very large number of transshipment services available. Singapore also serves as a consolidation hub for South East Asia ports to major markets such as the USA and Europe, thus there is a large increase in waiting time when these smaller services are not able to transship onto larger international liner services. Busan, Yokohama and Qingdao are ranked second, third and fourth after Singapore in terms of impact on waiting time when transshipment is disabled. This is likely due to their geographical location as gateway ports to the North China region, forming a bottleneck through which all services must pass. Tianjin has no impact on waiting time due to its position deep in the bay of Bohai. This means that services will not be routed via Tianjin as there will be some backtracking and hence time wasted.

The results provide an indication of the relative impact of these ports on the transportation network as a whole, which can be seen as a reflection of their connectivity. Further analysis can be performed using the same framework to obtain different insights, which can then be combined as part of the benchmarking process. For example, the results from the throughput model and the waiting time model can be integrated using the Pareto graph in Fig. 1 below.

The grouping of ports shows that Singapore clearly has the highest connectivity in terms of both impact on throughput and impact on waiting time. Busan, Yokohama, Qingdao and Shanghai form a second group of ports that have similar connectivity rankings, with some tradeoffs between throughput and waiting time within the group. Dalian has a large impact on throughput, but a small impact on waiting time, which pulls down its overall connectivity ranking.

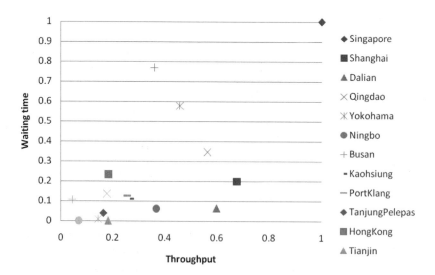

Fig. 1   Pareto analysis of throughput vs. waiting time.

3.3. *Impact of factors on individual ports*

The degree of complexity inherent in a port system can often make it very difficult to determine the influence that any single factor can have on the port's performance. For example, carriers and shippers can base their port selection on any number of factors such as cargo handling charges, reliability or land accessibility. The complexity is compounded by the fact that a given factor can play very different roles for different ports. In any benchmarking study, it is useful to examine each port's strengths and weaknesses, as well as the factors contributing to their success or failure.

In order to analyze the impact of these factors on individual ports, this section will utilize a network flow model (MCNFM) which is similar to the model presented in Lee *et al.*, (2006). The model attempts to derive optimal shipping decisions by balancing the actual cost of transportation and opportunity cost of transportation while satisfying constraints on supply and demand, port capacity and link capacity. Thus, the model is able to account for the trade-offs between transportation cost and shipping time. The model is based on several assumptions, which are presented below.

Assumptions:

(i) The volume of trade demand between countries is known and deterministic. Trade data are obtained in terms of dollar volume of trades, from the United Nations COMTRADE database.

(ii) Each port is assumed to be the sole sea link gateway for that country. For China, there are five ports in the study, namely Hong Kong, Shenzhen, Qingdao, Ningbo and Shanghai. As it is difficult to model demand and supply operations for 5 different ports under the same country, they are grouped according to their geographical proximity and linked into three regions. These are: Hong Kong-Shenzhen, for Southern China; Shanghai-Ningbo, for Middle China; and Qingdao, for Northern China.

(iii) For the segregation of Chinese ports and regions, it is assumed that the port for each specific region handles all of the demand and supply for the region. In effect, each region acts as a separate country with their own gateway ports.

(iv) As the study focuses on the connectivity of maritime ports and does not consider land links, each port-country/region pair will be considered as one node. For example, Kaohsiung-Taiwan will be considered as one node and Hong Kong-Shenzhen will be considered as another node serving the southern part of China.

(v) As per Lee *et al.* 's model, the rationality of shippers is assumed, and informed choices are made on the basis of perfect information.

Parameters:

N: Set of nodes (port-country/region)
O: Set of origin nodes
D: Set of destination nodes
L: Set of links
M: Set of ports
α_{ij}:   The freight cost of transporting one TEU of cargo from node i to node j
β_{ij}:   The time incurred of transporting one TEU of cargo from node i to node j
ρ: Nominal cost coefficient
λ: Average monetary value of one TEU for the single commodity
K_k: the capacity of port k
S_k^{od}:   Denotes the amount of supply or demand for the commodity at node k for origin-destination pair o-d. S_k^{od} is non-zero only if node k either the destination or the origin. If node k is the origin, then S_k^{od} is either 0 or negative (supply). If node k is the destination, then S_k^{od} is either 0 or positive (demand).

MCNFM:

$$\text{Min} \quad \sum_o \sum_d \sum_{ij} (\alpha_{ij} \cdot x_{ij}^{od}) + \lambda \cdot \rho \sum_o \sum_d \sum_{ij} (\beta_{ij} \cdot x_{ij}^{od}) \tag{4}$$

$$s.t. \quad \sum_{i \in N} x_{ik}^{od} - \sum_{j \in N} x_{kj}^{od} = S_k^{od}, \quad \forall o \in O, \quad d \in D, \quad k \in N, \tag{5}$$

$$\sum_o \sum_d x_{ij}^{od} \leq U_{ij}, \quad \forall(i,j) \in L, \tag{6}$$

$$\sum_o \sum_d \left(\sum_{i \in N} x_{ik}^{od} + \delta_k S_k^{od} \right) \leq K_k, \quad \forall k \in N \tag{7}$$

$$x_{ij}^{od} \geq 0, \tag{8}$$

where x_{ij}^{od} are the decision variables which indicate the number of container flow in TEU shipped from origin o to destination d through link (i,j). The objective value of the linear program minimizes the sum of two terms, of which the first is the cost of transportation from node i to j and the second is the opportunity cost of transporting goods from i to j. Equation (6) is the conservation of flow constraint, which states that the amount of goods going into port k for all origin — destination pairs must be equal to the sum of the demand or supply and the amount of goods going out of port k. Equation (7) states that the total amount of goods utilizing a particular link must be within the link capacity. Equation (8) limits the total goods passing through a port to within its annual capacity, where an index is adopted which indicates whether port k is a supply port or demand port, i.e.,

$$\delta_k = \begin{cases} 1, & \text{if port } k \text{ is a supply port,} \\ 0, & \text{otherwise.} \end{cases} \tag{9}$$

The last equation (9) sets the decision variables to be non-negative.

The model is implemented using ILog OPL Development Studio 5.5 based on a CPLEX solver, and generates throughput figures for all origin-destination pairs. By varying certain parameters such as link capacities and frequencies, we can analyze the interactions among the major Asia Pacific ports to help understand the competitive landscape in the region.

3.3.1. *Case study*

This subsection focuses on the analysis of the impact of some factors on the nine major ports mentioned in Subsection 3.1.1. In our analysis, each port is assumed to be the sole sea link gateway for that country. For example, Kaohsiung is assumed to be the only port serving Taiwan, although in actual fact Keelung is also a major port in Taiwan. Data from three major liner shipping companies was used in our analysis, namely Maersk, Evergreen and Mediterranean Shipping Company. Detailed data for other shipping companies can be added in future studies to obtain more comprehensive results. In our analysis, we focus on the major shipping partners with the Asia Pacific region: the United States of America and Europe. The model requires some calibration to ensure consistency with actual throughput figures as only selected ports, liner companies and regions are included in the study, which is based on trade data from 2002. However, the relative proportion of throughput between the ports is fairly accurate.

The model is implemented using ILog OPL Development Studio 5.5 based on a CPLEX solver. In total, it has 6571 decision variables and 826

Table 6   Comparison of model throughput and actual throughput.

Port	Throughout (MODEL)	% of total generated throughput	Throughput (ACTUAL)	% of total actual throughout
Singapore	28975954	19.51	29918200	19.45
Shanghai — Ningbo	39412267	26.53	39206000	25.49
Busan	21648917	14.57	13425000	8.73
Qingdao	5155761	3.47	10320000	6.71
Hong Kong - Shenzhen	35114351	23.64	45661888	29.69
Tanjung Pelepas	15000000	10.10	5600000	3.64
Kaohsiung	3241739	2.18	9676554	6.29
Total	14764783	100.00	153807642	100.00

linear constraints. Table 6 below compares the generated throughput from the model with actual port throughput data.

The overall generated throughput for the model is lower than the actual throughput. This is to be expected as only selected ports are included in the study, and thus the results are not exhaustive. Furthermore, regions such as Australia, Middle East, Africa and South America are not included in the study. However, the generated throughput is still relatively accurate, which can be attributed to the fact that Europe and the United States are the main trading partners of Asia-Pacific countries and thus form the majority of the container throughput for Asia-Pacific container ports. For the purposes of this study, which focuses on the topic of port competition, the absolute throughput is not as important as the relative proportion of throughput between the ports and also the proportionate changes in throughput when certain variables are changed. Ranking ports in terms of proportions, the relative magnitude of the throughput figures are accurate except for Tanjung Pelepas and Shanghai-Ningbo. This can be attributed to the aggregation of Hong Kong and Ningbo into a single port and also the approximation of certain data.

Moving towards the comparison of throughput at the individual port level, the throughputs of Singapore and Shanghai-Ningbo ports reflected as a proportion of the total are relatively close to the actual throughput, contributing a combined percentage of 46.04%, which is almost half of the overall throughput. All the other ports, with the exception of Qingdao, have throughput that is less than the actual throughput. This might be attributed to the same reason mentioned previously (not all trading partners considered) and the assumption that only a single port is used to represent each country. In actual fact there can be numerous ports in a single country, for example Tanjung Pelepas, Port Klang in Malaysia, and Keelung, Kaohsiung in Taiwan.

After developing a model that can adequately describe and approximate the actual container shipping system in Asia-Pacific, sensitivity analysis can be carried out to examine the impact of different factors on major ports. As an example of such analysis, we consider the impact of varying link capacity on Singapore port and Busan port. The rationale for choosing these two ports is that they are major container shipping hubs, Singapore being a global transshipment hub and Busan being a key transshipment hub for the North Asia region, particularly for the transpacific trade route serving North America. For these two transshipment hubs, link capacities are more important than other ports, as they transfer cargo more frequently than other ports. The ability to receive and send out large amount of cargo is essential to transshipment ports. Sensitivity analysis for these two transshipment hubs might provide insights on the impact on other ports in the region if their link capacities are altered.

Singapore port

Figure 2 shows the variation in throughput (measured in thousands of TEUs) of the various ports in the study when the overall link capacities of Singapore port are varied. Specifically, the cases where the link capacities are reduced to 50% and 30% are considered.

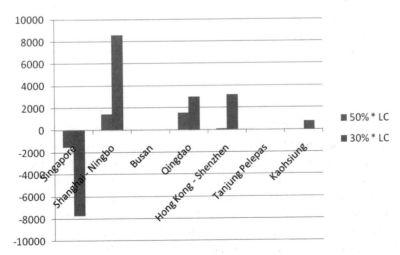

Fig. 2   Effects of reducing Singapore port's link capacities.

It was found that the traffic flow for Singapore takes a much bigger dip when the link capacity is reduced from 50% to 30%. This concurs with the observation that link capacities are very important to the success of a major transshipment hub such as Singapore. At 50% of the link capacity, we see increased traffic flow to Qingdao, Shanghai-Ningbo and Hong Kong-Shenzhen, in decreasing order.

When Singapore's link capacity is reduced to 30%, traffic flow increases greatly to Shanghai-Ningbo, Hong Kong-Shenzhen and Qingdao (in decreasing order), which all lie on the coastline of China. This is due to container flow that normally passes through Busan and Kaohsiung being first diverted through the three China ports due to insufficient capacity on the Singapore-Busan/Kaohsiung direct links with the reduction in Singapore's link capacity. These three ports can share the burden of a reduced capacity on Singapore's side. Shanghai-Ningbo's throughput experiences the greatest increase as it is geographically advantageous for both direct shipments to the USA or transshipment to USA via Busan (it is the next closest port to Busan while being nearer to Singapore than Qingdao). Both these possibilities can raise the throughput of Shanghai-Ningbo port by a large amount, leading to the observation from the model. This suggests Shanghai-Ningbo's strategic importance as a port in Asia Pacific, being well endowed with a superb geographical location which can provide alternatives for Singapore's USA-bound cargo which traditionally is shipped via Busan.

Tanjung Pelepas, being the closest port to Singapore in this study, surprisingly did not see any change in traffic when Singapore's link capacities are reduced. One possible reason for this is that the port is already operating at its maximum capacity in the model, which indicates an area for possible future refinement.

Busan port

The analysis is done on Busan in a similar manner, with link capacities reduced to 70% and 50%. The results are summarized in Fig. 3.

At 70% of the link capacity, increased traffic flow to almost all other ports can be observed, with the exception of Tanjung Pelepas. The ports of Shanghai-Ningbo, Hong Kong-Shenzhen, Kaohsiung and Qingdao, in decreasing order, see the greatest increases in traffic. These ports are within the geographical vicinity of Busan. This reinforces the previous findings that Shanghai-Ningbo is Busan's greatest competitor, and that a port's link capacity has an impact on neighboring ports. In addition, with the exception of Qingdao, the increase in throughput again seems to be related

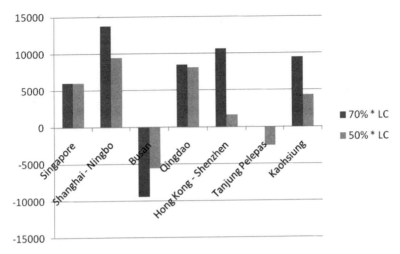

Fig. 3   Effects of reducing Busan port's link capacities.

to the geographical location. Shanghai-Ningbo is further north than Hong Kong-Shenzhen, which is roughly location in the same region as Kaohsiung. A preliminary hypothesis is that the nearer the port, the greater the gain in traffic of that neighbor port when the target port's link capacity is decreased. That is, port competition is localized.

In the 50% case, the above hypothesis is strengthened. With Singapore's throughput remaining at roughly the same levels, the ports with the greatest increase in throughput are Shanghai-Ningbo and Qingdao, which are the ports nearest to Busan. However, the overall improvement in neighboring ports is not as great compared to the 70% case. This might imply that as the link capacities for Busan port decreased, the overall throughput of the region is being hampered. This suggests inter-dependencies between ports in the same region, even though there is competition between them.

The sensitivity analysis performed on both Singapore and Busan reveal that Shanghai-Ningbo has the potential to be a major competitor in the region, at least when considering link capacities. Shanghai-Ningbo stands to gain the most from any reduction in link capacities through both Singapore and Busan and could become a key port for the Asia-Pacific region in the future.

4. Conclusion and Discussion

Port competition has grown more intense with the growth of international trade and global economies. Asia is the pre-eminent region in the world container throughput and the competition among Asian container ports is even more intense than in other regions of the world, highlighting the need for effective benchmarking methodologies. This paper focused on benchmarking major ports in the Asia-Pacific region from three aspects: port efficiency, port connectivity and impacts of factors on individual ports. Using DEA models, it was found that Singapore, Hong Kong and Shanghai rank as the most efficient. When considering a network framework to evaluate connectivity in terms of throughput capacity and waiting time, Singapore ranks as the most well connected, followed by Busan, Yokohama, Qingdao and Shanghai. Lastly, the sensitivity of ports towards the impact of various factors can be measured and benchmarked using a network flow model, which revealed the importance of Shanghai-Ningbo as major competitors to Singapore and Busan.

Some facts should be mentioned as follows:

- Most of the data used in the paper are collected from CI-online and some other resources. However, some of the data could not be found and they had to be estimated which will lead to inaccuracy of data. In addition, some data used may be outdated or inaccurate as there is no reliable universal source. So, further study may be required.
- Only nine ports and three liner shipping companies are included in this study for port efficiency and impact of factors on individual ports. The small number of ports and liner companies may not be representative of the whole maritime transportation network of Asia-Pacific region and this could likely lead to some inaccurate conclusions. More ports and liner shipping companies should be involved in future studies.
- Port connectivity in this paper is still a preliminary study and as mentioned earlier it is closely related to some other factors such as waiting time, responsiveness and cost. Thus, it will be more meaningful if we integrate these factors together in our connectivity analysis.

References

1. Barros, C.P. (2003). The measurement of efficiency of Portuguese sea port authorities with DEA. *International Journal of Transport Economics*, 30(3): 335–354.
2. Bemowski, M.T. (1991). The competitive benchmarking wagon. Quality Progress, January, pp. 11–16.
3. Coelli, T.J. (1996). A guide to DEAP version 2.1: a data envelopment analysis (computer) program, unpublished manuscript, University of New England at Armidale.

4. Cuadrado, M., Frasquet, M. and Cervera, A. (2004). Benchmarking the port services: a customer oriented proposal. Benchmarking: *An International Journal*, 11(3): 320–330.

5. Cullinane, K.P.B., Wang, T.F., Song, D.W. and Ji, P. (2006). The technical efficiency of container ports: comparing data envelopment analysis and stochastic frontier analysis. *Transportation Research A*, 40(4): 354–374.

6. Goss, R. (1990). Ecomomic policies and seaports: 4. Strategies for port authorities. *Maritime Policy and Management*, 17(4): 273–287.

7. Hoffmann, J. (2005). Liner shipping connectivity. UNCTAD Transport Newsletter, 27: 4–12.

8. Huybrechts, M., Meersman, H., Van De Voorde, E., Verebeke, A., Van Hooydonk, E. and Winkelmans, W. (2002). Port competitiveness, an economic and legal analysis of the factors determining the competitiveness of seaports, p. 18.

9. Kleywegt, A., Goh, M.L., Wu, G.Y. and Zhang, H.W. (2002). Competition between the Ports of Singapore and Malaysia. The Logistics Institute — Asia Pacific, National University of Singapore, Singapore.

10. Lee, L.H., Chew, E.P. and Lee, L.S. (2006). Multicommodity network flow model for Asia's container ports. *Maritime Policy & Management*, 33(4): 387–402.

11. Lirn, T.C., Thanopoulou, H.A., Beynon, M.J. and Beresford, A.K.C. (2004). An application of AHP on transhipment port selection: a global perspective. *Maritime Economics and Logistics*, 6: 70–91.

12. Liu, C.C. (2008). Evaluating the operational efficiency of major ports in the Asia-Pacific region using data envelopment analysis. *Applied Economics*, 40: 1737–1743.

13. Low, J.M.W., Lam, S.W. and Tang, L.C. (2009). Assessment of hub status among Asian ports from a network perspective. *Transportation Research Part A: Policy and Practice*, 43: 593–606.

14. Malchow, M. and Kanafani, A. (2004) A disaggregate analysis of port selection. *Transportation Research Part E: Logistics and Transportation Review*, 40: 317–337.

15. Marquez-Ramos, L., Martinez-Zarzoso, I., Perez-Garcia, E. and Wilmsmeier, G. (2006). Determinants of maritime transport costs. Importance of connectivity measures. Presented to International Trade and Logistics, Corporate Strategies and the Global Economy Congress Proceedings, September, Le Havre.

16. Martinez-Budria, E., Diaz-Armas, R., Navarro-Ibanez, M. and Ravelo-Mesa, T. (1999). A study of the efficiency of Spanish port authorities using data envelopment analysis, *International Journal of Transport Economics*, 26(2): 237–253.

17. McCalla, R., Slack, B. and Comtois, C. (2005). The Caribbean basin: adjusting to global trends in containerization. *Maritime Policy and Management*, 32(3): 245–261.

18. Murphy, P.R., Daley, J.M. and Dalenberg, D.R. (1992). Port selection criteria: an application of a transportation. *Logistics and Transportation Review*, 28: 237–254.

19. Notteboom, T. (2006). The time factor in liner shipping services. *Maritime Economics and Logistics*, 8: 19–39.

20. Panayides, P.M., Maxoulis, C.N., Wang, T.F. and Ng, K.Y.A. (2009). A critical analysis of DEA applications to seaport economic efficiency measurement. *Transport Reviews*, 29(2): 183–206.

21. Pardali, A. and Michalopoulos, V. (2008). Determining the position of container handling ports, using the benchmarking analysis: the case of the Port of Piraeus. *Maritime Policy & Management*, 35(3): 271–284.

22. Park, K.R. and De, P. (2004). An alternative approach to efficiency measurement of seaports. *Maritime Economics & Logistics*, 6(1): 53–69.

23. Poitras, G., Tongzon, J. and Li, H. (1996). Measuring port efficiency: an application of data envelopment analysis. Working paper (Singapore: National University of Singapore).

24. Roll, Y. and Hayuth, Y. (1993). Port performance comparison applying data envelopment analysis. *Maritime Policy & Management*, 20(2): 153–161.

25. Rugman, A.M. and Verbeke, A. (1993). How to operationalize Porter's diamond of international competitiveness. The International Executive, 35(4): 17–39.

26. Saaty, T.L. (1977). A scaling method for priorities in hierarchical structures. *Journal of Mathematical Psychology*, 15: 234–281.

27. Sharmaa, M.J. and Yu, S.J (2009). Performance based stratification and clustering for benchmarking of container terminals. *Expert Systems with Applications*, 36(3): 5016–5022.

28. Slack, B. (1985). Containerisation, inter-port competition and port selection. *Maritime Policy and Management*, 12(4): 293–303.

29. Song, D.W. and Yeo, K.T. (2004). A competitive Analysis of Chinese Container Ports using the Analytic Hierarchy Process. *Maritime Economics and Logistics*, 6: 34–52.

30. Tang, L.C., Low, J.M.W. and Lam, S.W. (2008). Understanding Port Choice Behavior¡[a] A Network Perspective. Networks and Spatial Economics, published online.

31. Tiwari, P., Itoh, H. and Dio, M. (2003). Shippers' port and carrier selection behaviour in China: a discrete choice analysis.*Maritime Economics and Logistics*, 5: 23–39.

32. Tongzon, J. (2001). Efficiency measurement of selected Australian and other international ports using data envelopment analysis. *Transportation Research A,* 35(2): 113–128.

33. Tongzon, J. and Heng, W. (2005). Port privatization, efficiency and competitiveness: some empirical evidence from container ports (terminals). *Transportation Research Part A: Policy and Practice*, 39: 405–424.

34. Ugboma, C.H., Ugboma, O. and Ogwude, I. (2006). An analytic hierarchy process (AHP) approach to port selection decisions — Empirical evidence from Nigerian ports. *Maritime Economics and Logistics*, 8: 251–266.

35. Valentine, V.F. and Gray, R. (2001). The measurement of port efficiency using data envelopment analysis, in: *Proceedings of the 9th World Conference on Transport Research*, 22–27 July (Seoul, South Korea).

36. Wilmsmeier, G., Hoffmann, J. and Sanchez, R.J. (2006). The impact of port characteristics on international maritime transport costs. *Research in Transportation Economics*, 16: 117–140.

37. Wilmsmeier, G. and Hoffmann, J. (2008). Liner shipping connectivity and port infrastructure as determinants of freight rates in the Caribbean. *Maritime Economics and Logistics*, 10: 130–151.

CHAPTER 5

IS PORT THROUGHPUT
A PORT OUTPUT?

Wayne K. Talley

Department of Economics, Old Dominion University,
Norfolk, Virginia, 23529 U.S.A.
wktalley@odu.edu

This paper investigates the question: is port throughput a port output in port economic production and cost functions? The paper concludes that the answer to this question is no. Port throughput is the amount of cargo received from carriers that passes through the port. A port does not produce throughput but rather provides interchange service for the cargo that it receives. By using port throughput as a measure of port output (or the dependent variable) in a port's interchange service economic production function: (1) the interchange service economic production function no longer exists and (2) the port's long-run total cost and short-run variable economic cost functions cannot be derived. A measure of port interchange service that is consistent with measures used heretofore by ports for evaluating their performance is the port throughput ratio — the ratio of cargo interchanged to the total time incurred in interchanging the cargo.

1. Introduction

Port cargo throughput is the amount of cargo that passes through a port from one transport carrier to another. The technical (or productive) efficiency of cargo ports has been investigated in the literature under the assumption that the cargo output of ports is cargo throughput. A port is technically efficient if its output is the maximum obtainable output in the use of given levels of resources. A port's economic production function represents this functional relationship. Port cargo throughput as a measure of port output is found in empirical port economic production studies by Chang,[1] Kim and Sachish[2] Bendall and Stent,[3] Dowd and Leschine[4] and Liu.[5] Container port output measured as TEU (twenty-foot equivalent unit) throughput has been used in empirical Data Envelopment Analysis (DEA) and stochastic frontier analysis models for investigating the relative technical efficiency among container ports by Tongzon,[6] Cullinane, Song, Ji, and Wang,[7] and Cullinane and Song.[8,q]

The cost efficiency of cargo ports has also been investigated in the literature under the assumption that the cargo output of ports is cargo throughput. A port is cost efficient if its output is provided at minimum costs. A port's economic cost function represents this functional relationship. Port cargo throughput as a measure of port output is found in empirical port economic cost studies by Kim and Sachish,[2] Martinez-Budria,[9] Jara-Diaz, Martinez-Budria, Cortes, and Vargas,[10] Jara-Diaz, Martinez-Budria, Cortes, and Basso[11] and Jara-Diaz, Tovar and Truillo.[12,r]

This paper investigates the question: is port throughput a port output in port economic production and cost functions? The paper concludes that the answer to this question is no. The output of ports is interchange service, i.e., ports use its resources to interchange cargo provided by carriers between arriving and departing carrier vessels and vehicles. The amount of cargo that is interchanged is the port's throughput. The paper also investigates the impact of measuring port output as port throughput rather than as port interchange service in the investigation of port technical and cost efficiencies.

In the next section, a port interchange service economic production function is presented, followed by discussions of port operating options and the port resource function in Section 3 and 4, respectively. In Section 5, the impacts of using port throughput as a measure of output in port economic production and cost functions are discussed. In Section 6, measures of port interchange service are presented. Finally, conclusions are found in Section 7.

2. Port Production Function

For the sake of simplicity, only the throughput of a container port is used in the following discussion. However, the discussion can be applied to any

---

[q]For further discussion of technical efficiency and ports, see Cheon, Dowall and Song18, Song, Cullinane and Roe17, Talley14, Talley15, Wang, Cullinane and Song19 and Yan, Sun and Liu.20

[r]A review of these cost studies is found in Tovar, Jara-Diaz and Trujillo.21 An estimate of an input distance function for investigating the technical inefficiency of ports, where tons of port cargo throughput are used as measures of port outputs, is found in Rodriguez-Alvarez, Tovar and Trujillo.22

type of port cargo. It is important to note that a port does not produce throughput, but provides interchange service to the cargo that it receives. For example, a container port does not produce TEUs (twenty-foot equivalent units) or 20 foot containers but rather provides interchange service for the TEUs that it receives, i.e., import TEUs received from ocean vessels are interchanged with domestic vessels and land vehicles and vice versus for export TEUs; transshipment TEUs are interchanged between vessels.

In order for a container port to provide TEU interchange service, at least two parties must be in agreement. If either party is not in agreement, no container port TEU interchange service will occur. Specifically, transportation carriers must be willing to transport containers to and from a port and the port must be willing to interchange containers that are received. If the port is unwilling to accept containers, even though carriers are willing to provide it with containers, no port TEU interchange service will occur. If carriers are unwilling to provide the port with containers, even though the port is willing to accept containers, no port TEU interchange service will occur. The port can not force carriers to provide it with containers and carriers can not force the port to accept its containers.[cs]

From the above discussion, a container port interchange service economic production function is a function that relates the maximum amount of TEU interchange service provided by the port to the levels of resources utilized in providing interchange service and the number of TEUs provided by carriers, i.e.,

$$\text{TEU Interchange Service} = f(R_i; \text{TEUs Provided by Carriers}) \quad i = 1, 2 \ldots M \qquad (1)$$

where, TEU interchange service is the service of interchanging TEUs between arriving and departing carrier vessels and vehicles at the port and R_i is the ith resource.

If the port adheres to its interchange service economic production function in the provision of TEU interchange service, the port is technically efficient in the provision of this service. Further, since the dependent variable of a firm's economic production function is the amount of output provided by the firm, it thus follows that the dependent variable of economic production function (1), TEU Interchange Service, is the amount of output provided by the container port.

---

[s]c. The exception is when the port and carrier are owned or leased by the same firm.

3. Port Operating Options

A firm that produces a tangible product seeks to produce homogenous units of the product, i.e., it does not seek to produce a product that varies in quality. Alternatively, a port that provides a service (rather than producing a tangible product) does seek to provide a service that varies in quality, e.g., to improve the quality of its interchange service in order to attract more cargo from carriers. The means by which a port can differentiate (or vary) the quality of its interchange service have been referred to in the literature as the port's operating options.[13,14,15]

Operating options for container ports, for example, include[14]: (1) loading and unloading service rates, i.e., TEUs loaded and unloaded per unit time to and from vessels, vehicles and port equipment; (2) entrance gate reliability (percent of time that the port's entrance gate is open for vehicles); (3) departure gate reliability (percent of time that the port's departure gate is open for vehicles); (4) harbor waterway reliability (percent of time that the port's harbor waterway is open to navigation); (5) berth reliability (percent of time that the port's berth is open to the berthing of vessels); (6) berth accessibility (percent of time that the port's berth adheres to authorized depth and width dimensions); (7) harbor waterway accessibility (percent of time that the port's harbor waterway adheres to authorized depth and width dimensions); (8) the monetary loss of cargo, vessel property, inland carrier property and port equipment property in port to theft; and (9) the monetary damage to cargo, vessels, inland carrier vehicles and port equipment in port.

4. Port Resource Function

A port resource function for the ith resource (R_i) relates the minimum amount of this resource to be employed by the port to the port's levels of operating options and the mount of cargo provided by carriers (see Talley[14]). If the cargo is TEUs, then the port resource function may be expressed as:

$$R_i = R_i(\text{OPTION}_1, \text{OPTION}_2, \text{OPTION}_n \ldots \text{OPTION}_N;$$

$$\text{TEUs Provided by Carriers}) \ i = 1, 2 \ldots M \qquad (2)$$

where, OPTION_n is the nth operating option of the port.

If no excess capacity exists for the port's ith resource, then a change in the port's nth operating option for the purpose of increasing the quality of its interchange service will require that an additional amount of the ith resource be employed by the port. If an increase in the quality of service requires an increase in the nth operating option, e.g., an increase in the vessel loading service rate, an increase in R_i will also occur, i.e., $\partial R_i/\partial \text{OPTION}_n > 0$. If an increase in the quality of service requires a decrease in the operating option, e.g., a decrease in the lost of port cargo to theft, an increase in R_i will also occur, i.e., $\partial R_i/\partial \text{OPTION}_n < 0$. If no excess capacity exists for resource R_i, an increase in TEUs Provided by Carriers to the port will result in an increase in the amount of resource R_i used by the port, i.e., $\partial R_i/\partial$ TEUs Provided by Carriers > 0.

5. Container Port Output

Rather than using TEU interchange service as the amount of output provided by a container port, the literature heretofore has instead used TEU throughput. However, what are the impacts in doing so on the estimation of container port's economic production and cost functions for investigating container port technical and cost inefficiencies?

5.1. *TEU Throughput and the port production function*

Theoretical economic production functions utilized heretofore in the estimation (via econometric and frontier analysis procedures) of port economic production functions are those that have appeared in the literature in the estimation of economic production functions of firms that produce a tangible product rather than provide a service. Specifically, these production functions relate the maximum amount of a product (or output) produced by a firm to the levels of resources utilized, i.e.,

$$\text{Output} = g1(Ri) \quad i = 1, 2, \ldots M \qquad (3)$$

Port studies where estimates of economic production function (3) are found include, for example, studies by Chang,[1] Bendall and Stent,[3] Liu,[5] Dowd and Leschine[4] and Cullinane and Song.[8] In the estimation of economic production function (3) for container ports, output is measured as TEU Throughput.

Criticisms in using function (3), where output is measured as TEU Throughput, to investigate the technical efficiency of container ports include: a) container ports provide a service as opposed to producing a tangible product, b) economic production function (3) unlike interchange service economic production function (1) does not capture the fact that

the number of TEUs received by ports are not determined by the ports
but by carriers through their port-choice decision making process and c)
interchange service economic production function (1) no longer exists if
interchange service is measured as TEU Throughput (see Proposition 1
below).

Proposition 1: *Interchange service economic production function (1) no
longer exists if interchange service is measured as TEU Throughput.*

Proof. If TEU Interchange Service in function (1) were replaced with
TEU Throughput, a problem would arise, i.e., from the fact that TEU
Throughput = TEUs Provided by Carriers[t] or TEUs interchanged by the
port are the same TEUs provided to the port by carriers. Port output
(a dependent variable) cannot be a function of itself and the function be
a port economic production function. A function by definition denotes the
relationship between a dependent variable and one or more independent
variables. A function does not denote the relationship between a dependent
variable and a set of variables that includes the same dependent variable.
Alternatively, a variable cannot be both a dependent and an independent
variable in the same relationship and the relationship be a function. ∎

5.2. *TEU Throughput and port cost functions*

A port will be cost efficient in the provision of its output if it provides
its output at the least cost, given the prices that it pays for the resources
employed. The relationship between the least or minimum costs and the
amount of output provided is the port's economic cost function. In order to
be cost efficient, it is necessary that the port be technically efficient in the
provision of this output. Otherwise the port could provide the same amount
of output with at least a lesser amount of one resource and therefore incur
lower cost in providing this output (for given resource prices).

5.2.1. *Long-run total cost function*

The long-run is a time period that is sufficiently long enough so that
amounts of all resources employed can be varied. To be cost efficient in

---

[t]In practice, i.e., once a time period is specified (say a year), this equality may not hold
exactly. For example, containers that are received by a port at the end of the year may
not depart from the port (and therefore become port throughput) until several days into
the next year.

providing output in the long-run time period, a port will minimize its long-run total cost (LTC) in the provision of technically efficient output (represented by the interchange service economic production function (1)), i.e., minimize LTC subject to interchange service economic production function (1). Further, LTC will be minimized subject to resource function (2), since the latter represents the minimum amount of a given resource to be utilized the port given the amount of TEUs provided by carriers and the levels of the port's operating options. LTC is the sum of the products of the amounts of resources (R_i) utilized by the port and the prices (P_i) incurred in their utilization, where P_i is the price incurred by the port for the ith resource.

From the above discussion, the port's long-run total cost function will be derived by:

Minimizing LTC $= \sum P_i R_i$

subject to

TEU Interchange Service $= f(R_i; \text{TEUs Provided by Carriers})$ (4)

$R_i = R_i(\text{OPTION}_1, \text{OPTION}_2, \text{OPTION}_n \ldots \text{OPTION}_N;$

TEUs Provided by Carriers) $\quad i = 1, 2, \ldots M$

The choice variables in the optimization of equation (4) are the port's operating options. That is to say, the port would select the levels of its operating options (or quality of service). This selection in turn determines, given the TEUs provided by carriers, the port's minimum levels of resources to be employed — which are also the levels to be employed that minimize the port's long-run costs (LTC). From this optimization, the port's long-run economic total cost function can also be derived, where LTC is a function of the prices of the resources employed by the port, the port's TEU Interchange Service and TEUs Provided by Carriers, i.e.,

LTC $= \text{LTC}(P_i, \text{TEU Interchange Service},$

TEUs Provided by Carriers) $\quad i = 1, 2, 3 \ldots M$ (5)

Proposition 2: *Port long-run economic total cost function (5) cannot be derived if TEU Interchange Service is measured by TEU Throughput.*

Proof. By Proposition 1, interchange service economic production function (1) does not exist if interchange service is measured as TEU

Throughput. Hence, Eq. (4) cannot be optimized to obtain long-run economic total cost function (5).[u] ∎

5.2.2. Short-run variable cost function

To be cost efficient in providing output in the short-run (a time period that is sufficiently short so that the amounts of one or more resources employed cannot be varied, i.e., are fixed), a port will minimize its short-run variable cost (SRVC) subject to interchange service economic production function (1), R_M amount of fixed resource M and resource function (2), i.e.,

Minimize SRVC $= \sum P_i R_i$

subject to

TEU Interchange Service $= f(R_i; R_M$, TEUs Provided by Carriers) (6)

$R_i = R_i(\text{OPTION}_1, \text{OPTION}_2, \text{OPTION}_n \dots \text{OPTION}_N;$

TEUs Provided by Carriers) $\quad i = 1, 2 \dots M - 1$

As for Eq. (4), the choice variables in the optimization of Eq. (6) are the port's operating options. From the optimization, the port's short-run economic variable cost function can be derived, where SRVC is a function of the prices of the variable resources (1 through $M - 1$) employed by the port, the port's TEU Interchange Service, TEUs Provided by Carriers and R_M amount of fixed resource M, i.e.,

SRVC $= $ SRVC$(P_i,$ TEU Interchange Service,

TEUs Provided by Carriers, $R_M) \quad i = 1, 2, 3 \dots M - 1$ (7)

---

[u]Critics of the use of port aggregated cargo throughput (i.e., the aggregation of the throughput of two or more different types of cargo) as a measure of port output in the estimation of port economic production and cost functions note that port aggregated cargo throughput generates aggregation-bias estimates of technical and cost efficiencies for ports that handle more than one type of cargo, e.g., solid bulk, liquid bulk, general non-containerized and containerized cargoes. To avoid aggregation bias in such studies, Kim and Sachish[2] recommend that the port throughput of each type of cargo be used as port output measures rather than using a single aggregate throughput measure. In a port cost function estimation study by Martinez-Budria,[9] aggregate throughput is used as the measure of port output. However, the study makes the restrictive assumption that the cost share of a ton of any type of cargo is independent of the activity where it is handled.

Proposition 3: *Port short-run economic variable cost function (7) cannot be derived if TEU Interchange Service is measured by TEU Throughput.*

Proof. By Proposition 1, interchange service economic production function (1) does not exist if interchange service is measured as TEU Throughput. Hence, Eq. (6) cannot be optimized to obtain short-run economic variable cost function (7).   ∎

5.2.3. *Port cost function estimates*

Estimates of long-run economic total cost functions for various types of ports are found in studies by Kim and Sachish,[2] Martinez-Budria,[9] Jara-Diaz, Martinez-Budria, Cortes and Vargas,[10] Jara-Diaz, Martinez-Budria, Cortes and Basso[11] and Jara-Diaz, Tovar and Trujillo.[12] Port long-run total costs in these studies were expressed as functions of port throughput and the prices of the resources employed by ports — but not functions of port.

Interchange service.[v] Measures of port throughput found in these studies include tons of: cargo, container general cargo, non-container general cargo, liquid bulk cargo, dry bulk cargo and roll-on roll-off (Ro-Ro) cargo.

As noted in Proposition 1, a port's TEU throughput is also the TEUs provided by carriers to the port. This is also true for any type of port throughput — i.e., a port's throughput of any type is also the amount of cargo of any type provided by carriers to the port (to be interchanged between arriving and departing carrier vessels and vehicles). Hence, the port long-run economic total cost functions used in the above cost studies may be alternatively stated as functions, where port long-run economic total cost is expressed as a function of the amount of cargo provided by carriers to the port and the prices of the resources employed by the port. However, the derived long-run economic total cost function for a container port (function 5) has long-run total cost being a function of not only of TEUs provided by carriers and the prices of resources employed, but also of the port's TEU interchange service. For a port handling any type of cargo, function (5) would be rewritten as port long-run total cost being a function of cargo provided by carriers, prices of resources employed and the level of cargo interchange service.

---

[v]The one possible exception is the study by Jara-Diaz, Martinez-Budria, Cortes and Basso[11] that includes the explanatory variable, index added of other activities that use part of the port infrastructure.

Since the above port cost studies do not include the explanatory variable, cargo interchange service, in obtaining their long-run total cost function estimates, the studies' estimated parameters are statistically biased due to specification error, i.e., from a relevant explanatory variable having been omitted from the estimation (see Pindyck and Rubinfeld.[16,w] The studies conclude that ports exhibit economies of scale, i.e., port long-run total cost does not increase in proportion to the increase in port throughput. Would this conclusion hold if the explanatory variable, cargo interchange service, had been used in obtaining the long-run total cost function estimates?

6. Port Interchange Service Measures

6.1. *Port revenue*

In a study by Song, Cullinane and Roe,[17] a container port's output was initially described as an interchange service. Specifically, the output was defined as the container turnover derived from the delivery of container terminal services and measured by the revenue received for these services. However, problems arose in the use of revenue as a measure of container port output. Port revenue not related to interchange service, e.g., from property sales, had to be excluded. The wide diversity of accounting systems used by port terminals in the study resulted in an intractable problem of separating out revenue attributable to different sources. In the end, the study chose instead the output measure of annual container throughput TEUs.[2]

Even if revenue received from port interchange service could be measured without error, port revenue would not be an appropriate measure of port interchange service. That is to say, port revenue from port interchange service and the amount of port interchange service provided are not necessarily positively related. Port revenue may be increasing, while port interchange service is decreasing over time. Port revenue is the product of port price and port interchange service. Port revenue will increase over time even though the amount of port interchange service demanded is decreasing over time if the demand for port interchange service is price inelastic, all else held constant.

---

[w]A parameter estimate is biased when the expected value of the estimated parameter does not equal to the population parameter. Since the bias will not disappear as the size of the sample increases, the parameter estimate is also inconsistent.

6.2. *Port throughput ratio*

A feasible measure of port interchange service is the port throughput ratio — i.e., the ratio of port throughput to the time that the throughput is in port. This ratio would be determined by dividing the amount of cargo provided by carriers to and passing through a port (or interchanged) by the total time that the cargo is in port. This total time is a measure of the utilization of resources by a port in the interchange of cargo. Note that if a port's throughput is increasing faster over time than the time it incurs in interchanging throughput, the port's throughput ratio will be increasing over time. Alternatively, it can be stated that the port's technical inefficiency is decreasing over time, since the port's utilization of resources is not increasing in proportion to the increase throughput.

The throughput ratio is consistent with measures used heretofore by ports for evaluating the performance of their interchange service, e.g., vessel loading and unloading services rates — the amount of cargo loaded to and from vessels per unit of time. Also, ports seek to reduce the dwell time of cargo, e.g., storage time per TEU stored or increase the reciprocal, TEUs stored per unit of storage time.

7. Conclusion

This paper investigated the question: is port throughput a port output in port economic production and cost functions? The paper concludes that the answer to this question is no.

A port provides interchange service for cargoes provided by transport carriers, i.e., a port uses its resources to interchange cargoes between arriving and departing carrier vessels and vehicles. A port's interchange service economic production function relates the maximum amount of cargo interchange service provided by the port to the levels of resources utilized in providing interchange service and the amount of cargo provided by carriers. The amount of cargo that is interchanged is the port's throughput.

A port's long-run economic total cost function relates the minimum long-run total cost to be incurred by the port to the prices of the resources employed by the port, the port's cargo interchange service and the cargo provided by carriers. A port's short-run economic variable cost function relates the minimum short-run variable cost to be incurred by the port to the prices of the variable resources employed by the port, the port's cargo interchange service, the cargo provided by carriers and the amount(s) of the fixed resource(s).

By using port throughput as a measure of port output (or the dependent variable) in a port's interchange service economic production function the: (1) interchange service economic production function no longer exists and (2) port's long-run economic total cost and short-run economic variable cost functions cannot be derived. A measure of port interchange service that is consistent with measures used heretofore by ports for evaluating the performance of their interchange service is the port throughput ratio — the ratio of cargo interchanged (or throughput) to the total time incurred in interchanging the cargo.

References

1. Chang, S. (1978). *Maritime Policy and Management*, 5.
2. Kim, M. and Sachish, A. (1986). *Journal of Industrial Economics*, 35.
3. Bendall, H. and Stent, A. (1987). *Maritime Policy and Management*, 14.
4. Dowd, T. and Leschine, T. (1990). *Maritime Policy and Management*, 17.
5. Liu, Z. (1995). *Journal of Transport Economics and Policy*, 29.
6. Tongzon, J. (2001). *Transportation Research Part A*, 34.
7. Cullinane, K., Song, D.-W., Ji, P. and Wang, T.-F. (2004). In Review of Network Economics (Special Issue), Ed. Talley, W.K.
8. Cullinane, K. and Song, D.-W. (2006). In Port Economics: Research in Transportation Economics, Eds. Cullinane, K. and Talley, W.K. (Elsevier, Amsterdam).
9. Martinez-Budria, E. (1996). *Revista Asturiana de Economia*, 6.
10. Jara-Diaz, S., Martinez-Budria, E., Cortes, C. and Vargas, A. (1997). 25th European Transport Forum: Proceedings Seminar L (PTRC, London).
11. Jara-Diaz, S., Martinez-Budria, E., Cortes, C. and Basso, L. (2002). *Transportation*, 29.
12. Jara-Diaz, S., Tovar, B. and Trujillo, L. (2005). *Transportation*, 32.
13. Talley, W.K. (1994). *Logistics and Transportation Review*, 30.
14. Talley, W.K. (2006). In Port Economics: Research in Transportation Economics, Eds. Cullinane, K. and Talley, W.K. (Elsevier, Amsterdam).
15. Talley, W.K. (2009). Port Economics (Routledge, Abingdon, United Kingdom).
16. Pindyck, R.S. and Rubinfeld, D.L. (1981). Econometric Models and EconomicForecasts (McGraw-Hill Book Company, New York).
17. Song, D.-W., Cullinane, K. and Roe, M. (2001). The Productive Efficiency of Container Terminals: An Application to Korea and the UK. (Ashgate, London).
18. Cheon, S., Dowall, D.E. and Song, D.-W. Transportation Research Part E, (forthcoming).
19. Wang, T.-F., Cullinane, K. and Song, D.-W. (2005). Container Port Production and Economic Efficiency (Palgrave-Macmillan, Basingstoke).

20. Yan, J., Sun, S. and Liu, J. (2009). *Transportation Research Part B*, 43.
21. Tovar, B., Jara-Diaz, S. and Trujillo, L. (2007). *Maritime Policy and Management*, 34.
22. Rodriguez-Alvarez, A., Tovar, B. and Trujillo, L. (2007). *International Journal of Production Economics*, 109.

CHAPTER 6

A FRAMEWORK FOR MODELLING AND BENCHMARKING MARITIME CLUSTERS: AN APPLICATION TO THE MARITIME CLUSTER OF PIRAEUS

Vassilios K. Zagkas and Dimitrios V. Lyridis

National Technical University of Athens School
of Naval Architecture & Marine Engineering
Laboratory for Maritime Transport, Athens, Greece
vzagkas@naval.ntua.gr
dsvlr@central.ntua.gr

This paper investigates the factors that contribute to the decisions of companies from key maritime sectors to be established in a specific area that evolves into a network of firms. It is also the scope of this paper to investigate and benchmark the circumstances under which a network of firms is transformed into a competitive Maritime Cluster. In this framework we present methods for developing and evaluating possible models for the Cluster creation and development, addressing more specifically the case of Piraeus. The concentration of the research on the Greater Area of Piraeus as opposed to the whole country is considered given that the country's major port is located in the Greater Area of Piraeus which constitutes a very active maritime community. Furthermore, the paper will give a short introduction into new computational methods such as Agent Based Modelling for simulating the networking process within maritime clusters and managing their life cycle. This will give an insight of firm survival strategies within the cluster, optimum timing for new entrants in the cluster and overall cluster management.

1. Introduction

This paper addresses the role of regional entrepreneurial networks and their evolution into dynamic cluster formations through the emergence of competitive advantages. Several theories have been applied in the study of clusters; such theories are agglomeration economics, industrial districts, spatial economics, and economic geography- all of them being useful tools. However, the competitiveness theory as developed by Michael Porter in the 1990's is the most well-known theory on clusters and their economic behavior. The integration of Porter's theory with the maritime context can give a pragmatic approach to Maritime Clusters. Planning and structuring the maritime cluster can be considered as a cyclical process consisting of iterative cycles infatuated by governmental or private initiatives. However, managing the maritime cluster, retaining and enhancing its competitive advantages in the context of international competition are complex issues related to dynamic systems and complexity theory. In the framework of this research a sophisticated computational model such as Agent Based Modelling is employed in order to simulate the actions and interactions of firms that act as autonomous individuals in the maritime cluster, with a view to assessing their effects on the cluster system as a whole.

2. Conceptual Definition: Cluster Theory and Maritime Clusters

The increasing interest in cluster structures and their valuable outcomes have led us into their investigation and tempted us into their application on the Maritime sector. The development of clusters is by some seen as the only way to overcome the risk of being outperformed in the global economy (Lagendijk, 2000). Cluster theory was first introduced almost a century ago by Alfred Marshall under the term of industrial districts. In Principles of Economics he described the phenomenon as '*the concentration of specialized industries in particular localities*' (Marshall, 1922). The concept of knowledge spill-over and externalities were crucial elements on Marshall's theory, as he elegantly states: '*The mysteries of the trade become no mysteries, but are as it were in the air*' (Marshall, 1922). Industrial districts enjoy the same economies of scale that only giant companies normally get. Specialized suppliers arrive. Skilled workers know where to come to ply their trade. And everyone involved benefits from the spill-over of specialized knowledge (Surowiecki, 2000). Later on, the competitiveness theory as developed by Michael Porter presented in his 1990's book, 'The Competitive Advantage of Nations', is the most well known theory on clusters and their economic behavior. Economic sciences hesitate to get involved in subjects where the use of numbers and quantities is limited. However, thanks to Porters approach, it became widely known that cluster has a very good impact on the economy and many papers like this one struggle to define how much good that is. The main argument in Porter's theory is that firms and not nations compete in international markets and the presence of competing clusters is a key dynamic factor to nation competitiveness. Porter's Diamond can be better comprehended through the crucial question: '*Why do firms based in particular nations achieve international success in distinct segments and industries?*' (Porter, 1990). According to Porter the answer lies in four broad attributes of a nation that shape the environment in which local firms compete and promote competitive advantage. Those four elements are: Factor Conditions, Demand Conditions, Related and Supporting Industries, Firm strategy, structure and rivalry (Porter, 1990). The integration of the four elements in Porter's theory with the maritime context can give a pragmatic approach to Maritime Clusters. However, it is very difficult to speak eloquently about the cluster of firms or the competitive advantage of some regions or cities without explicitly taking into consideration the 'space'. Over the years, economists have neglected spatial issues, due to the difficulty of modeling increasing returns and imperfect competition. Thus, the study of economic geography and space was pushed to the periphery of economic theories (Krugman, 1991).

This overview shows that there is an old and strong theoretical background for clusters addressing both their economic and spatial matters, originating from industrial districts and they are related to into agglomeration economics, spatial economics and the competitiveness theory. This paper addresses the need for a theory integrating the four factors of Porter's theory with economic values and spatial development; a system that will reveal the effect the micro level has on the macro level — the whole cluster.

3. The Concept of Maritime Clusters

Experience around the world has shown that the concept of clustering suits particularly well to maritime businesses. There are numerous benefits, ranging from specialized labor to targeted training, from increased market awareness to connections with R&D institutes and from strategic co-operations to inter-related maritime activities (Wijnolst, 2009). Despite the large maritime industry in Europe and worldwide, we have little systematic information concerning the degree of interaction between maritime firms. The European network of maritime clusters is one of the pioneering initiatives concerning the cross-country maritime cluster of Europe. Several country reports, as Norway's' and Netherlands', have been published there revealing the structure and some quantitative data of their maritime cluster (Wijnolst, 2006).

The need for a flexible theory to base our research on, has directed us towards a bottom-up approach to the maritime cluster concept. Therefore, the first task was the conceptual definition of the maritime cluster, hence, maritime cluster as per se in our research can be defined as: '*The outcome of one or more spatial consolidations, of cooperating — competing firms and institutions within all sectors, sub-sectors and economic activities directly or indirectly linked to the shipping industry, maritime transport and generally the utilization of the sea*'.

Based on the above definition, it is then necessary to define the sector and sub-sectors that make up a maritime cluster. The European commission has identified the following traditional maritime sectors in Europe (E.C. Report, 2009), as shown in Table 1.

However, many differences exist per country and maritime cluster regarding the scope of the maritime industry and its specialization. The European Network of maritime clusters give us a more narrow or more pragmatic perspective on the sectors of European maritime cluster. Here, we have eight sectors: Shipping, Shipbuilding, Marine equipment, Seaports,

Table 1   Traditional maritime sector according to EC study (E.C. Report, 2009).

Traditional Maritime Sectors (EC study)	
Shipping	Scrapping
Shipbuilding	Offshore supply
Ports & Related Services	Cable & Submarine telecom
Classification Societies	Inland Shipping
Repair & conversion	Naval Shipbuilding
R&D and Education	Dredging & Maritime works
Equipment Manufacturing	Recreational Vessels
Support Services	Fishing & Aquaculture

Maritime services, Yacht building, Offshore services and Fishing (E.C. Report, 2009).

Much of the literature on clusters has ignored the issue of market structure. On the other hand, literature on maritime clusters is dedicated on replicating the market networking structure that exists without investigating in depth the reasoning for such networking. In the framework of approaching the concept of maritime clusters, there is the need to analyze the structure of the network and identify the key relationships that control the supply and demand in the maritime sector. The study on the maritime service sector in London (Grammenos, 1992) gave us an insight on how different firms in the shipping industry are interconnected. The study came to a model with a core centre of three main sectors: Charterers, Owners and Brokers around whom ancillary services revolve. The model proposed here is based on one fundamental value that shall govern the behavior of the cluster; there is supply and demand of knowledge between the different categories that live in the cluster and among the firms that populate each nod in a micro perspective. In practice, the demand and supply of knowledge can be translated into exchange of services and goods among the firms. The triangle of charterers, owners and brokers (Grammenos, 1992) has been replaced by one triangle incorporating Ship-owners, Ship-managers and Charterers and another inverted triangle included in the previous one, presenting brokers as the intermediary activity between supply and demand. The following scheme presents the idea of the *double inverted triangle*.

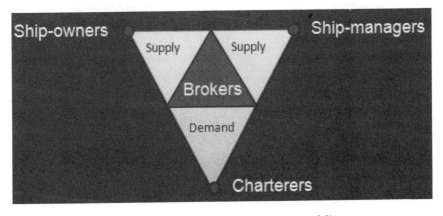

Fig. 1   Double inverted triangle, market modeling.

As Fig. 1 above implies, the demand for shipping is expressed in the market by charterers who seek vessels for the movement of their cargoes while on the other side the supply is expressed by Ship-owners and Ship-managers who offer ships to charterers for their needs. The most usual case is that the supply side and the demand side employ brokers to match their needs. This is the reason why brokers absorb demand information from Charterers with the one side of the triangle and supply information from Ship-owners and Ship-managers with the other two sides respectively. The purpose of this model is only to present the core of shipping activity around which satellite services revolve and they constitute as a whole a universe of maritime activities — the maritime cluster. For the moment, this model is simplified and relieved from complexity issues that certainly exist in the market. Further on, the addition of satellite services around the core activity creates a network that can be considered as a maritime cluster, shown in Fig. 2 below. In order to harmonize the above model with the definition of maritime clusters we should also consider the factor of localization. It is therefore essential to enhance the model with the spatial dimension, meaning that the players of the core activity with the firms that offer the satellite services are co-locating in a region that can be therefore characterized as a Maritime Cluster.

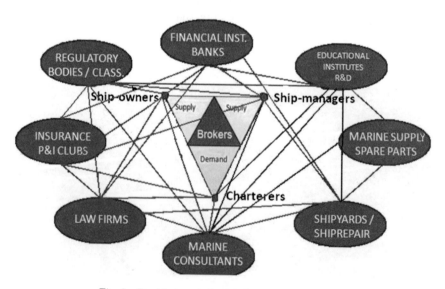

Fig. 2   Double inverted triangle, market modeling.

4. Spatial Paradigm: The Greater Area of Piraeus

The greater area of Piraeus is the shelter of a significant number of companies and organisations that participate in the maritime activity of the country. This significant number can be roughly estimated at over 1.000 shipping companies, a vast number of technical companies, many banks with shipping departments and more than 1.200 marine services companies. Greek shipping has been considered among the most successful industries in the world accounting for about 4,392 ships, a figure translated in 8.7% of the world fleet being controlled by Greek owners (GSCC, 2008). Having in mind that Greece is one of the small countries in the world with a population of around 11 million people and ranking 96th in total area out of 231 countries worldwide (Wikipedia, 2009), some important observations arise. Such observations lead us to assume that there is a high density of maritime related services in Greece and that there shall be significant factors that create competitive advantages for the Greek shipping industry. The high density of maritime activity in a small country has directed our theoretical framework towards spatial theory. Mapping maritime related businesses in Greece has resulted in identifying four key regions that include the vast majority of firms with maritime activity. The most significant region that will also serve as our case study is the Greater Area of Piraeus, consisting of the city of Piraeus and seven of its adjacent suburbs. The other areas mentioned in line of importance are the Northern Suburbs of Athens, the Southern Suburbs of Athens and the City of Athens.

The cluster population of the greater area of Piraeus can be analyzed with the use of firm statistics created from the 'Greek-Cypriot Maritime Guide' (MIS, 2008). All registered firms are included in this dataset. Some of the firms in the dataset are members of larger groups or subsidiaries created for economic reasons. Therefore, the number of 'real' firms is overestimated. This shortcoming is not important for the purposes of this study, because the general picture of Piraeus's cluster is fairly reliable. The figure below identifies the importance of each area by the number of companies in each key sector for the maritime cluster.

The sector of maritime services in the greater area of Piraeus can be characterized as large and dynamic. It can compete with other key exporters of maritime services as London, whilst lacking in maturity and organization. The sector demonstrates cluster behaviour with geographic concentrations and interconnected companies, while the cluster forces that shall hold it together, seem weak and vague. It is vital that these forces be strengthened

V. K. Zagkas and D. V. Lyridis

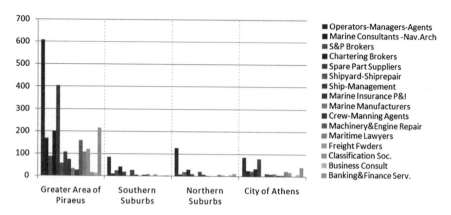

Chart 1   Sector fragmentation by region.

in the face of international competitive pressures that will try to pull the cluster apart. The axis of world economic activity is moving eastwards and competing centres in the Far East are expected to gain in stature (Lagendijk, 2000).

4.1. *The structure of the Piraeus maritime cluster*

The maritime industry in the greater area of Piraeus constitutes a complete cluster. It is composed of three main bodies: shipping, maritime services and maritime industry. The cluster is also surrounded by research & educational institutions, governmental bodies, the port & port authority and some maritime associations.

Figure 3 suggests that the three core segments in the cluster consisting of services that are directly connected to each other. This network of services is not abandoned in the marketplace, but it acts in the framework of big co-operating institutions that are concerned with the quality and the quality of the services provided. The maritime associations, port authority, research &educational institutions and governmental bodies not only are part of the cluster but they also surround it since they can contribute into policy making. The most important competitive advantages of the cluster are described and explained in textbox 1 below. Over the years, a variety of factors has affected the structure of the cluster as described above. However, there are four significant variables recognized: the agglomeration effects, internal competition, cluster barriers and heterogeneity (Langen, 2004). These variables will be later on specifically discussed for our case study.

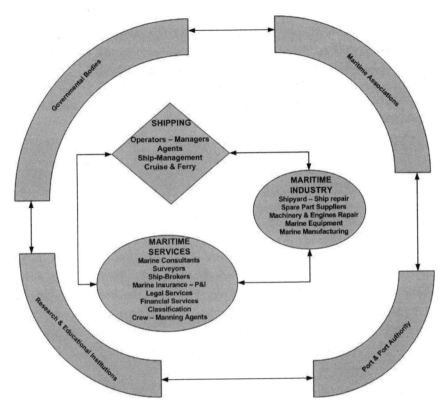

Fig. 3   Structure of the Piraeus cluster.

Textbox 1. The core segments of the Piraeus maritime cluster.

Shipping: Shipping is the core of this cluster and it is constituted by owners and operators of all kinds of vessels, e.g., bulk carriers, oil tankers, container ships, general cargo, gas carriers, reefer ships, fishing vessels, cruise ships and ferries. The shipping segment is considered to be the most important in the cluster not only because it is the largest network of companies, but also because shipping companies are the most international and fundamental ones in the internationalisation of the cluster (Wijnolst, 2009). Shipping companies create the excessive demand in services of high quality, hence stimulating innovation and creativity in the whole cluster. According to our survey 2009, there are 608 shipping firms in the greater area of Piraeus.

(Continued)

Textbox 1. (*Continued*)

Shipbrokers: The role of shipbrokers as explained earlier is crucial for the shipping market and for the cluster. The greater area of Piraeus hosts approximately 290 companies, composed from small, medium even large firms with international reputation and branch offices in important maritime centres.

Marine Consultants (Naval Architects — Surveyors): There is a massive concentration of technical offices or individual brand firms specializing in technical consultancy, ship design and surveying in Piraeus. According to our survey, there are 168 marine consulting firms active in the greater area of Piraeus, mostly addressing the demand created by shipping companies located in the area.

Spare Part Suppliers: Firms specializing in spare part supplies dealing with repairs in the shipping industry of Piraeus. More than 400 firms support the most demanding fleet of the world constantly.

Machinery & Engine Repairs: This segment is constituted of 160 companies, specializing in low cost repairs of machinery and engines; it is a crucial service for the shipping companies located in the area.

Legal Services: There are a large number of lawyers specializing in maritime law and consulting in the greater area of Piraeus, this segment consists of both big firms and individual lawyers, counting over 100 lawyers in the core area of Piraeus.

Banking & Financial Services: Another strong segment of the cluster. There are over 210 institutions, banks and firms specializing in financial services for the maritime sector in Piraeus. This includes local banks and firms as well as representatives of famous international institutions.

4.2. *The economic footprint of the maritime industry in the region*

There are several ways to assess the economic importance of an industry. Such are employment, profitability, productivity and knowledge externalities. The maritime industry in Greece is large, internationally competitive and geographically concentrated. These characteristics make it a very important asset of the Greek economy. The geographical concentration of the industry, as indicated above, tempts us to assess its economic footprint on the corresponding region.

The shipping sector, only one segment of the maritime cluster contributes strongly to the Greek Economy. Only for the year 2007, the net income from shipping was $17 billion, meaning 7% of the GDP covering 28% of the trade balance deficit (World Bank, 2007). The added value of the maritime sector in Greece according to a report form '*Policy Research Corporation*' is €6400 million, that is 3.24% of the GDP of the country. *Added Value* is the net output of a sector after adding up all outputs and subtracting intermediate inputs.

Concerning the *employment* factor, the maritime industry in Greece is a substantial employer. There around 76,200 people employed in the sectors of the maritime cluster. The concentration of the cluster in Attica and Piraeus represents 43.3% of the total maritime employment in Attica and 55% of that in the Greater Area of Piraeus.

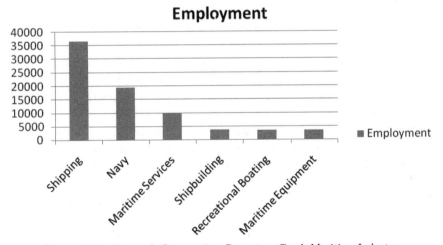

Source: Policy Research Corporation, Report on Greek Maritime Industry.

4.3. *SWOT analysis*

The SWOT analysis is a summary of the results coming from a preliminary survey undertaken as well as a comparison with other prominent clusters and a review on the perspective of sector experts. On one hand, strengths (S) and weaknesses (W) can reveal the internal conditions of the cluster and its current position while on the other hand, opportunities (O) and threats (T) focus on future growth and suggestions (Wijnolst, 2009).

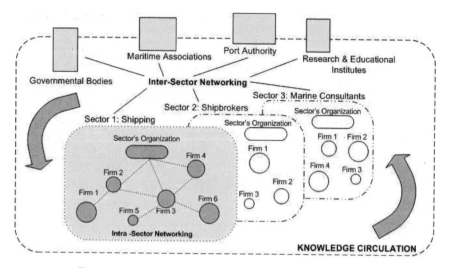

Fig. 4 Multi-scale cluster organization for ABMS modeling.

- There may certainly be additional weaknesses, threats, etc. other than those mentioned in Figure 4 such as:
- Deindustrialization of Greece and gradual relocation of industries from Western Europe to emerging countries with cheaper labor force.
- Relatively expensive Greek labor force compared to emerging countries wage levels.
- The Mediterranean Sea no longer plays a central role in international trade as it used to, since it is faraway from BRIC countries.
- Major shipyards are nowadays located in the Far East, very faraway from Greece.
- Countries such as China are beginning to develop commercial fleet as well.
- The new international industrial center is in the Far East.
- London is Europe's major trade center and represents direct threat to Piraeus, as it attracts so many Greek shipping companies.

Therefore Table 2 must not be considered as comprehensive.

The basic conclusion from the analysis is that Greece is a leading maritime center, but there are a number of actions and initiatives that should take place in order to become a maritime centre of excellence, providing quality maritime services for international customers. The identified opportunities should be encountered in policy strategies and the threats and weaknesses shall be balanced out by direct public and private initiatives.

Table 2   Summarized SWOT analysis based on survey & interviews.

SWOT-Strengths
Large number of leading international companies in shipping
The largest fleet of the World
Concentration of diversified maritime services at a specific region
Very high gross earning from shipping & high added value from the entire maritime
    industry
Strong networking, trust and family ties between the member of the maritime
    community
Competitive taxation for shipping companies
Concentrated knowledge and experience in maritime matters
International reputation and long maritime tradition

SWOT-Weaknesses
Declining Seafarer's labor force
Lack of R&D and Private Maritime Institutions
Limited number of institutions for maritime education
Lack of awareness of the significance of the maritime industry in the Greek Economy
Lack of Maritime Cluster Organization backed from public and private initiatives
Lack of regulatory framework and initiatives for high capaciy activities such as: Marine
    Arbitration, Greek P&I Clubs, Greek Stock Market open to shipping, New financing
    tools for shipping

SWOT-Opportunities
Attract more shipping companies and other maritime activities in Greece
Greater government efforts to promote and support Greek maritime interests
    internationally
Further support and development strategies for the small enterprises of the Greek
    maritime community
Creation of an official Greek Maritime Cluster organization
Support the shipbuilding industry and increase shipbuilding capacity
Increase Research & Development expenditures on maritime matters
Greater focus on quality of maritime services and international competition
Increase networking and intitiatives for collaboration

SWOT-Threats
Competition from countries with high innovation index and advanced R&D activity in
    the maritime sector
Insufficient flow of skilled labor into the sector
Lack of integrated maritime policy and strategic plan
Lack of unified safety culture and environmental awareness

5. Methods for Evaluating and Benchmarking
Maritime Clusters

There is an increasing request for a more fact-based input to Cluster
formulation and development. Good practice methods have not yet been
identified in evaluating and benchmarking clusters since the field is still
considered to be relatively new. One of the most recent and robust
methodologies in the literature is the one developed by the European

Cluster Observatory (www.clusterobservatory.eu). This European initiative has created a Cluster Mapping database that is built in the intersection of regions and sectors in Europe. The European Cluster Observatory has developed the so-called 3-Star evaluation method to determine the strength of clusters. This method utilizes three factors which are the following:

Size of the Cluster (represented by employment share in relation to total European employment),

Specialization (represented by the proportion of employment in a cluster category over total employment in the region),

Focus (represented by the share of cluster category in the regional economy). Further to the classification of cluster strength the European Cluster Observatory uses two additional cluster performance indicators: innovation index (based on the based on the Regional Innovation Scoreboard 2006) and world export share.

Another very important factor in the development and performance of clusters is the human capital. Many researchers in the area of economic growth believe that the growth of human capital is the most important factor in developing knowledge-based economies and innovative processes. Padmore and Gibson also emphasize the role of human capital in the development of clusters. Their GEM model (Groundings-Enterprises-Markets) includes factors very similar to Porter's diamond (Padmore *et al.*, 1998). The GEM model was created to assess the potency and effectiveness of a given cluster by scoring each of six determinants from the following major categories: supply determinants, structural determinants, market determinants.

There is a wide variety of performance indicators that can be utilized for the benchmarking of entrepreneurial systems; however for our research on maritime clusters we have divided them into general cluster sizing indicators and intra-cluster indicators. The general cluster sizing indicators are predominately used to exhibit the existence of the cluster in the area under study and also to rank the cluster in comparison with other competing ones in terms of size and influence on the local and national economy. However the general cluster sizing indicators, such as the size of the cluster in terms of the number of firms or the percentage share in GDP are not only useful for determining the size of the cluster or for ranking it against others, they are also essential in order to determine whether the cluster has reached the specialized critical mass to develop positive spill-overs and linkages. On the other hand intra-cluster indicators aim to reveal the performance of internal cluster dynamics that contribute to the existence and the development of the cluster.

Starting with the general cluster sizing indicators we have excluded some possible general performance indicators for being simplistic and non-proper for the aim of our research. These are: productivity and foreign Direct Investment. Productivity refers to metrics and measures of output from production processes, per unit of input. It is therefore a non-applicable measure for dynamic — evolving systems as clusters, since it does not facilitate measures for capturing the changes of cluster networking population (Maillat, 1998). Last, Foreign Direct Investment is not sufficient since clusters are dominated by local players focusing on outward investments; investments that can be measured by added value and employment indicators. On the other hand, inward investments can be more helpful on performance but still insufficient.

After identifying and reviewing the indicators that are not satisfactory for our study, the general performance indicators, below, are analyzed that are more suitable for maritime clusters.

5.1. *General cluster sizing indicators*

The dynamics of clusters depend strongly on structural, economic and social performance indicators, such as:

Cluster Structure Indicators: The structure of the cluster is a fundamental element of cluster's strength. The type and number of maritime sectors that exist in the cluster play a significant role. The broader is the cluster in terms of sectors, the greater is the probability of high networking and cluster effects. There is also a distinction between sectors in the cluster. Not all sectors have the same importance; for example, shipping is the most important sector, highly contributing to the added value generated and also pulling the demand for services within the cluster. Therefore, maritime sectors are attributed with weights of importance, thus representing their effect on the cluster in a more realistic way.

Another structure indicator is the population of the cluster and its sectors. It is critical to enter population barriers for sectors in order to qualify as members of the cluster. It is believed that a linear relationship exists between the population of firms in the cluster and the strength of the cluster.

Economic Performance Indicators: The use of standard economic performance indicators as used in all markets and economies is reasonable since these indicators can be used as tools for benchmarking and comparison against other clusters. Therefore, the following indicators in Textbox 2 are suggested for measuring the economic performance of maritime clusters.

Textbox 2. Economic performance indicators for maritime clusters.

Direct/Indirect Added Value: Added Value refers to the additional value of a commodity over the cost of commodities used to produce it from the previous stage of production. For the delivery of maritime services, the value added consists mainly of labor expenses, depreciation and profit before tax. Since this indicator cannot be directly calculated, for this research, added value is directly linked to employment data and the added value per person statistic from Eurostat (Policy Research 2008).

% Share in GDP: Gross domestic product (GDP) is defined as the "value of all final goods and services produced in a country in one year; it is more simply the total output of a region. Therefore the share that the Cluster has in the total output of the country is strongly indicating the importance of clustering, especially if it monitored over a time series, while the cluster matures.

Growth Rate: Economic growth is the increase in the amount of the goods and services produced by the Cluster over time. This can be conventionally measured for our purposed as the percent rate increase in share in the GDP.

Employment: This is the most stable and significant indicator for the performance of any business activity. Existing employment data are assessed and their correlation with the cluster performance and its added value are evaluated.

Risk Tolerance: It is a measure of how much a company will risk in order to gain a specified return. This paper strongly supports the use of risk tolerance as performance indicator for the cluster since we expect it to be proportional with the level of clustering and geographical concentration. Clustering is exposed to the effect of risk aversion, hence seeking greater returns that can substantially increase the firms' perceived utility.

(*Continued*)

Textbox 2. (*Continued*)

Profitability: Profitability can be problematic for measuring the performance of the cluster since clustering does not necessarily lead to higher profits of firms in the cluster (Langen, 2004). However all firms are ultimately driven by profitability and Clusters that their firms gradually improve the profitability can seen to be stronger and more stable. This indicator shows the efficiency of the company's activity.

5.2. *Intra-cluster performance indicators*

The amount and quality of knowledge circulating and spilling over between firms located in a cluster is dependent upon the cluster's size (which is determined by the indicators described above) the cluster structure and the intra-cluster dynamics. The later are processes as well as expected outcomes of the clustering effect. The main intra-cluster indicators are innovation (which can be measured with many different variants and perspectives), knowledge diffusion, intra-firm linkages and intra-cluster initiatives (new entrants-firms, institutions, collaboration schemes).

Innovation & Research Indicators: Innovation is a key factor to determine productivity growth. The importance of the cluster's structure is also present here. The existence of strong maritime services and marine equipment sectors indicates increased research activity and innovative spirit. According to many scholars, the more innovative the individual sectors are, the stronger the cluster becomes as a whole. Furthermore, the existence of leader firms significantly drives the innovation cycles within sectors, since they can signal demand and lead SMEs to an integrated research strategy. For the purpose of our research, we have defined a simple Innovation Efficiency Index (IEI) that is defined as the ratio of innovation outputs over innovation inputs. Inputs and outputs concerning innovation are the sum of specific sub-categories as defined in Table 3.

Calculating the innovation efficiency index is a challenging effort. However, there are several difficulties that arise from the use of the above factors. The above data are very demanding due to their rarity. The numbers used are based on existing quantitative databases, as Eurostat, reinforced by qualitative data received from questionnaires and interviews from sector experts. Producing the index requires that all data are normalized and then summed up, in order to calculate the desired index. The construction of a synthetic index requires comparability of data (Hollanders, Esser, 2007). The innovations indicators are incommensurate with each other as several

Table 3   Maritime cluster innovation inputs and outputs. Relevant data retrieved from (Hollanders, Esser, 2007).

INNOVATION INPUTS IN THE MARITME CLUSTER
Education Attainment Level
(% share of population in the Maritime Cluster aged 20–28 with upper secondary education in Maritime matters)

Participation in Maritime Seminars and Education
(% share of Maritime Cluster population)

Population having attained Msc or Phd on Maritime Education
(% share of Maritime Cluster population)

Public R&D expenditure for Maritime matters
(% share of Clusters' Added Value)

Maritime Businesses R&D expenditure for development
(% share of Clusters' Added Value)

R&D expenditures for high technology & manufacturing
(% share of Clusters' Added Value)

Firms that have developed in-house R&D and Innovation
(% share of Maritime Cluster firm population)

Firms co-operating in Innovative products and R&D
(% share of Maritime Cluster firm population)

Total Expenditures in Innovation for the Maritime Sector
(% share of Clusters' Added Value)

INNOVATION OUTPUTS FROM THE MARITME CLUSTER
Employment in high-tech services
(% share of total workforce in the Maritime Cluster)

Sales of new-to-market products
(% share of Clusters' Added Value)

Sales of new B2B products
(% share of Clusters' Added Value)

Number of scientific publications
(Measured per number of educated employees in the Maritime Cluster)

Number of patents
(Measured per number of employees in the Maritime Cluster)

New entries of companies
(% share of firm population in the Cluster measure over a time span of 5 years)

Firms that have developed in-house R&D and Innovation
(% share of Maritime Cluster firm population)

---

of them have different units of measurement. R&D expenditure indicators e.g., are expressed as a percentage of value added in Maritime Cluster while other indicators are expressed as share in population of firms or workforce. There are a number of normalization methods available. In this research we predict that the use of the two most common methods, *standardization (or z-scores*) and *re-scaling*, shall be the most suitable. The Innovation Index in therefore is computed as a weighted sum of its normalized component indicators:

Where Q is the number of innovation indicators.

$$\text{IEI} = \frac{\sum_{q=1}^{Q} [w_q^I(\text{out}_q)]}{\sum_{q=1}^{Q} [w_q^I(\text{in}_q)]}$$

Knowledge Diffusion: This is the process of communicating research, innovations and or knowledge to individuals, groups, firms or organizations. When firms of the same or adjacent sector are located in clusters they share a common set of values and knowledge. The significance of such values and knowledge form a cultural environment in which these firms co-exist. The nature of this cultural environment can either encourage or discourage the exchanged of knowledge and problems. The existing literature (Schrader, 1991) suggests that knowledge diffusion through informal channels happens as information trading. This refers to informal exchange of information between employees working for different and even competing firms. The exchanged knowledge can either be specific and of high value or general and of low value. To attain data on knowledge diffusion of the case study cluster we have used interviews and questionnaires. These are designed to obtain information on the knowledge base of the firms and also information on the value of transferred knowledge to and from collocating firms. Preliminary results of our above research demonstrate that leader firms of the cluster tend to have more developed knowledge base and stronger knowledge flow with other leader firms in the cluster. After normalizing and retrieving all data from interviews and questionnaires the pattern of knowledge diffusion will also be verified by a simulation with the aid of agent based modeling which is described in section 4 of this paper. The basic idea behind the knowledge diffusion simulation is that a sample population of the cluster is represented by nodes; each node is of equal importance and has knowledge to supply to one or more of the other nodes. As the simulation runs the weight of importance of each node changes in relation to the knowledge demand and knowledge supply functions attached to it.

5.3. *On the use of data and analysis for measuring performance of maritime clusters*

Numerous international studies have presented results which indicate that clusters have a positive impact on innovation and economic growth, therefore many organizations and even countries and regions have embraced the concept of clusters and try to develop them through specific initiatives. Facing the need to back-up the analysis of the maritime cluster of Piraeus, a range of tools was employed after being adapted to the needs of a maritime cluster. The methodology as described below is used in order to tackle the problem of cluster performance analysis regarding our selected case study.

Cluster Mapping: The first step of this project was the mapping of the cluster. In a very practical way, mapping of the cluster is classifying the firms by sector and marking them on the map. The patterns of clustering are thus very easy to identify. This is visual evidence that the cluster exists. Geographical proximity is then proved by numbers for each sector region and sub-region. This task is essential in cluster modeling in order to understand the practical structure of the maritime community and identify the regions of competence.

Cluster Database: The following step is the creation of a cluster database. The database carries information for nearly all the firms in the cluster. The maritime cluster of Piraeus consists of around 3,000 firms diversified in various maritime activities. A variety of data is contained, as location, number of employees, number of ships and number of new-building orders (in the case of shipping companies), annual turnover (where available), market share in the sector, patents, publications, education of employees, average wage etc. Those data are normalized and statistically analyzed in order to be used as performance indicators for the cluster as a whole.

Survey/Interviews: The most comprehensive tool, that can shed light into the dark corners of firms networking patterns, is the use of survey and much more the use of interviews with sector experts. For the purposes of our research, the second step after identifying cluster sectors is to identify the leading firms in the cluster for every sector and proceed with identifying the experts. All experts are invited to an interview so as to collect personal opinions about the structure of the maritime community in Piraeus. Except from firm experts' interviews applied to experts from organizations, classification societies, governmental bodies and educational institutes are interviewed as well. The results of the survey and interviews used as qualitative data are processed by SPSS software in order to identify correlation between key factors that drive the development of the cluster. The outcome is then used to feed the computational model and it is analyzed in the following section. The use of surveys and interviews is efficient for the localized maritime cluster of Piraeus, however it would be very difficult for these methods in order to apply multi-cluster comparison. Nevertheless, surveys and interview are considered an essential tool when studying and modeling a cluster, since it is vital to identify the consciousness of the major stakeholders.

Concluding cluster analysis can be achieved with several tools and enables accurate and effective policy and management intervention. It is essential to have a good understanding of a cluster's internal workings — components, structures, processes, routines and development pathways.

6. Computational Methods for Simulation and Life-Cycle Management of Maritime Clusters

Managing the maritime cluster, retaining and enhancing its competitive advantages in the context of international competition are complex matters that lend themselves to dynamic systems and complexity theory. In the framework of this research, a sophisticated computational model such as Agent Based Modelling is employed in order to simulate the actions and interactions of firms that act as autonomous individuals in the maritime cluster, with a view to assessing their effects on the cluster system as a whole. The model intends to simulate the simultaneous operations of multiple agents-firms, in an attempt to re-create and predict the actions of complex phenomena such as the maritime business environment. Agent–Based modeling has connections to many other fields; its historical roots can be traced in the study of complex systems (CAS) and has thereon extended into techniques and theories such as Cellular Automata, Swarm Intelligence, Network Science and Social Simulation.

6.1. *Agent-based modeling and simulation*

The increasing complexity of the world and its systems, calls for management tools that must be able to capture the whole lattice of their complexity. Industrial and governmental organizations frequently base their research and decision–making on fine data organized in the form of analytical databases. However, there are no robust tools for revealing emerging patterns from this data. Competitive advantage can be missed without the use of sophisticated tools. Simulation and Agent–Based modeling can contribute into assembling patterns from the chaotic interactions of firms. Agent–Based modeling is used to increase the capabilities of experts to grasp micro-level behavior and to relate this behavior to macro-level outcomes (North, Macal, 2007). This technique is based on the notion that unique rules, parts and components of a system are represented in the form of individual agents. Agents have varying influence and none of them can solely determine the ultimate outcome of the system. On the other hand, every agent contributes to the results in some way.

Implementing computational agents is the next step. Agents are the decision–making components in complex adaptive systems. They are attributed with sets of rules or behavior patterns that allow them to receive information, process them and then reflect them in the environment. Another characteristic of agent-through information processing is adaption and learning. Before modeling agents, it is important to understand their structure as units. Agents are individuals with a set of attributes and behavioral characteristics. These are explained in textbox 3.

Textbox 3. Carrying characteristics of agents.

Agent Attributes: There are various agent attributes. These are essentially some key characteristics of the agents that are ascribed by the user, in order to measure the outcome of the simulation. In an agent-based simulation, attributes are carried by each agent and can evolve or change over time as a function of each agent's learning experiences.

Agent Behaviors: Agents have behavior features that can vary from agent to agent in order to reflect pragmatic situations. There are two levels of rules. The first level specifies how the agent will react to routine events and the second level provides rules for the adaption of changing routines. Generally, agent behaviors follow three steps: 1. Agents evaluate their current state and determine their actions, 2. Agents execute the actions that they have chosen 3. Agents evaluate the results of their actions and adjust their rules.

In this case the firms within the cluster are agents. When seeking a detailed simulation the result is a multi-scale model of cluster behavior with the smaller scale firm interactions combined to produce the larger-scale activities of the cluster as a whole.

6.2. *Modeling case study: the maritime cluster of Piraeus*

The complex nature of our research, has directed us towards computer simulation with the use of an agent-based model. After identifying previously some characteristic of agent-based model, we need to define our model for the maritime cluster of Piraeus. Firstly, some principal assumptions need to be considered. Agents in this model represent only firms, organizations and institutions. Every agent belongs to a sector in the maritime cluster. For the purposes of the simulation, the population of firms in each sector is in not full scale; a sample of companies is attributed to each sector. The problem that this model addresses is to determine the emergence of competitive advantages for each individual firm within a cluster. In a knowledge based economy the source of competitive advantage for firms is no more limited to cost and differentiation advantages but it is linked to resources-competences that firms possess and their capability to create knowledge (Carbonara N *et al.*, 2006). The model seeks to investigate if the emergence of knowledge externalities drives the development of clusters and determines the factors that control some critical performance indicators for clusters.

All agents-firms have attributes and behaviors. These can change over time and by sector. After a detailed survey on sector experts, here are the selected attributes and behaviors for our model (Table 4).

Explaining each attribute, **Size:** The size of each firm is measured in accordance with the number of employees that have attained an educational degree, **Knowledge:** This is the heart of the model. As explained before, the long-term growth of firms and regions depend on their ability to continually develop and produce innovative products and services that are directly linked to knowledge. Services that are provided and acquired in the market are here modeled as demand and supply of knowledge. Knowledge is therefore exchanged within the cluster, with different rate of accumulation for its firm. Measuring the accumulation of that knowledge can present the emergence of competitive advantage in firms, **Innovation:** Is critical to measure the innovative capacity of each firm and sector. This is a derivative

Table 4   Attribute and behaviors of each agent–firm.

Firm	
Attributes	Behaviors
Size	Knowledge Demand
Knowledge Stock	Knowledge Supply
Innovation	Learning
Growth Rate	Moving in new positions
Risk Tolerance	
Market Share Targets	
Position on the grid	

of knowledge as described above, **Position:** This attribute indicated the position of the firm in a dimensional grid. The grid contains all the firms and the agent by calculating the maximization of his competitive advantage that depends on the knowledge stock and market share he can acquire; takes the decision to move or not on a more competitive position in the grid. The rest of the attributes are described before as performance indicators, that when attached to each agent they can derive valuable information. The first experimental stage of the simulation uses a sample population of firms from all sectors, assuming that they all position in a dimensional grid, having all the same knowledge capacity but different weight; something that depends on the firms' size. Starting the simulation, knowledge is circulated according to demand and supply. Then, firms try to locate where networking favors their competitive advantage, from this routine geographical concentrations arise and clusters of firms are developed. The results from this simulation are then validated against realistic data from the existing structure of the maritime community, in Piraeus. This confirms that the assumption of the initial model was pragmatic, that indeed, in reality, knowledge externalities drive clustering and that clustering of firms maximizes the performance indicators chosen. A multi-scale cluster model, as perceived, is shown below, with firms as subagents, sectors and relating institutions and bodies that are agents as well.

6.3. *Agent-based modeling toolkit*

There are a number of toolkits available for implementing agent-based modeling. Thanks to substantial public and private research, many computational environments have been developed and are now available for business use without any charge. The software environment for this research project is Repast (the REcursive Porous Agent Simulation Toolkit) and it is a leading open-source large scale ABMS toolkit. Repast was developed in order to support the development of extremely flexible models of agents focusing on social and economic simulation (North *et al.*, 2007). Repast's goal is to represent agents as discrete entities that act as social actors and are mutually defined with recombinant motives. The broader scope of the toolkit is to replay cases with altered assumptions (ROAD, 2004).

7. Conclusions

The traditional dynamic of Greek shipping companies and services that accompany them, together with special circumstances, contribute into making our era a unique opportunity for strengthening the development of the Piraeus & Greater Area maritime cluster. This emerging competitive advantage of the region must be nourished and encouraged.

Nowadays, there are significant opportunities to defend the existing Greek Maritime cluster formation and organise it, against both cost pressures and competition. However, in order to utilise such opportunities, it is essential that all stakeholders act with collective response on a cluster level basis. Talking about stakeholders, it is essential to identify them and assign their role and response to the cluster movement. According to the subject research, one of the major stakeholders is the Public sector and more specifically the Central and Local Government. Results from other cluster surveys have shown that the public sector has a major role in cluster formations. In fact, a supportive government is one of the most important criteria for the competitiveness of the cluster. Central government must develop enhanced understanding of the cluster and offer increased priority and support. This is also implemented in the agent-based model.

The awakening of the private sector is also essential. The behaviour of the private sector in Greece, as we know it today, must significantly change. Companies shall incorporate in their strategies the managerial theory of the 20th century. Cooperation amongst companies is a must for improved competitiveness and collective behaviour. Companies can be more efficient by developing a cross-selling culture in order to grow business across the cluster as a whole. All stakeholders shall develop a philosophy of partnership. The public and the private sector shall learn to work in the framework of a strong funded cluster organisation, pursuing the promotion of Piraeus as a global maritime services centre. Cluster initiatives and projects shall be pursued, both by the government and companies. The maritime identity of Piraeus shall be promoted worldwide and it should develop an image of offering cost-effective office space for smaller firms and associations, and opportunities for co-location to maximise cluster factors. Synergies shall be exploited with other services clusters.

Concluding, the efforts of the central government in these first critical steps of cluster development shall be based on supporting research and projects around the cluster and its built up. The results of this research should then form the basis for structuring public policies and financial proposals, as tax relaxations and land use for services localisation, which will favour the emergence of Piraeus as a global maritime services centre.

References

1. Carbonara, N., Albino, V. and Giannoccaro, I. (2006). Knowledge Externalities in geographical clusters: An agent — based simulation study. EIASM Workshop on Complexity and Management, Oxford.
2. European Commission Report (2009). DG Fisheries and Maritime Affairs studies: "Employment trends in all sectors related to the sea or using sea resources" and "Employment in the fisheries sector", http://ec.europa.eu/maritimeaffairs/study_employment_en.html, Accessed on 06/02/09.
3. Fisher Associates, (2004). 'The Future of London's Maritime services cluster: A call for Action', www.fisherassoc.co.uk, London.
4. Greek — Cypriot Maritime Guide, Marine Information Systems, 2009.
5. Greek Shipping Cooperation Committee/Lloyd's Register — Fairplay, February 2008 'Greek Controlled fleet annually by number of ships'.
6. Hollanders, H. and Esser, F.C. (2007). Measuring innovation efficiency INNO-Metrics Thematic Paper, PRO INNO Europe.
7. International Standard Industrial Classification, United Nations, 2009, http://unstats.un.org
8. Krugman, P. (1991). Geography and Trade, The MIT Press, Oxford.
9. Langen, P. (2004). The Performance of Seaport Clusters; a framework to analyze cluster performance and an application to the seaport clusters in Durban, Rotterdam, and the lower Mississippi, Erasmus University Rotterdam.
10. Lagendijk, A. (2000). Learning in Non-core regions: Towards 'Intelligent Clusters': Addressing Business and Regional needs. In Boekerna, F., Morgan, K., Bakkers, S. and Rutten, R. (Eds.). Knowledge, Innovation and Economic Growth: The Theories and Practice of Learning Regions (pp. 165–191), Edward Elgar, Cheltenham.
11. Marshall, A. (1922). Principles of Economics, London.
12. North, M.J. and Macal, C.M. (2007). Managing Business Complexity, Discovering Strategic lutions with Agent — Based Modeling and Simulation. Oxford Press, New York.
13. Porter, M. (1990). Competitive Advantage of Nations. MacMillan, London.
14. ROAD (2004). Repast homepage: http://repast.sourceforge.net/ (Accessed on 08/02/09).
15. Wijnolst, N. (Ed.) (2006). Dynamic European maritime clusters, Rotterdam: Dutch Maritime Network.
16. Wijnolst, N. and Wergeland, T. (2009). Shipping Innovation, IOS Press, Amsterdam.
17. Wikipedia (2009). http://en.wikipedia.org/wiki/List_of_countries_and_outlying_territories_by_total_ara (Accessed on 09/02/09).
18. World Bank Organization, Data & Statistics, http://web.worldbank.org/ (Accessed on 10/02/09).
19. Padmore, T. and Gibson H. (1998). Modelling systems of innovation: II. A framework forindustrial cluster analysis in region, Research Policy 26.
20. Schrader, S. (1991). Informal technology transfer between firms: cooperation through informal trading. Research Policy 20: 153–70.

CHAPTER 7

A PERFORMANCE EVALUATION STRATEGY TOWARDS DEALERS IN THE AUTOMOTIVE SUPPLY CHAIN

Min Chen, Wei Yan and Weijian Mi

Logistics Engineering School, Shanghai Maritime University
1550 Pudong Avenue, Shanghai 200135, P.R. China
minchen@shmtu.edu.cn

Owing to the paradigm shift from the make-to-stock (MTS) to the make-to-order (MTO), it was imperative to integrate the front-end market information for automotive supply chain. As a linkage between manufacturers and consumers, the performance of automobile dealers secured a crucial role in the automotive market-place. Accordingly in this study, a novel performance evaluation strategy has been developed for the automotive supply chain regarding four dimensional criteria, i.e., the financial condition, customer satisfaction, internal processes and self-innovation. More specifically, the balanced scorecard method was initially applied to evaluate the dealers' performance. Subsequently, a survey was conducted for the next-step evaluation. Consequently, the analytic network process (ANP) technique was employed to analyze the surveyed data. To this end, this strategy assisted automobile dealers in achieving both short-term and long-term objectives. Compared with traditional performance evaluation strategies, this approach could eliminate such disadvantages as time delaying and benefit orientation.

1. Introduction

In China, automobile manufacturers primarily rely on dealers to sell their automobiles and parts as well as to provide after-sale services to customers. This forms a partnership between the manufacturers and dealers. Although the dealers in China operate the 4S processes in a similar way like those in the Western, their self-fulfilled scale spanning from management capability, sale power to public relationship is still lacking. In the domestic automobile sales system, the vital player in the automotive supply chain is the manufacturers rather than other partners. As a result, they should balance and control regional market, establish comprehensive and in-depth sales network through which they secure the brand reputation and capitalize the profit (Zhang, 2003).

Pertaining to tougher competitions in today's market-place, together with additional pressures from current financial crisis, the paradigm of automotive industry was shifted from make-to-stock (MTS) to make-to-order (MTO). This could reduce the inventory and cost among the automotive supply chain. Based on this notion, it was imperative to integrate the front-end market information for automotive supply chain (Ma, 2007). As a linkage between manufacturers and consumers, the performance of automobile dealers secured a crucial position in the automotive market-place. As such, the performance evaluation became critical towards dealers (Dreyer, 2000; Chen *et al.*, 2008).

2. Problems of Dealer Evaluation

Dealers secure a crucial position between manufacturers and end customers for selling the products and providing customer services. Figs. 1 and 2 represent the sales and services process in the automobile supply chain. In particular, the 'dealer market information integration' refers to an integration of the 'sales' and 'services' functions in a supply chain, whereas

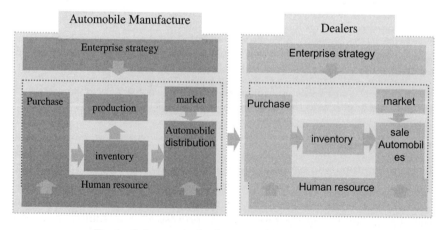

Fig. 1   Sales process in the automobile supply chain.

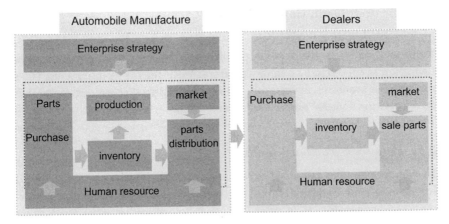

Fig. 2   Services process in the automotive supply chain.

the dealers' performance evaluation by manufacturers is aimed at enhancing their sales and service capabilities.

Currently, the performance evaluation mainly focuses on the accounting results of dealers, considering fairly about longer-term competence improvement of dealers. This impairs the automobile supply chain as a whole. There exist several impacts on the dealers' performance evaluation. In details,

1. The dealers' performance evaluation was conventionally emphasized on financial indices, thus these indices tended to be time-delaying and could not dynamically reflect the market information;
2. The internal business processes were usually lacking of evaluation and could not be objectively assessed on the supply chain operation;
3. The self-competence development of dealers was not well addressed because the competitive edge of entire supply chain was rarely considered.

These might result in following consequences.

1. Only focusing on the sales quantity rather than the potential dealers' development and royalty;
2. Inaccurately reporting on client order number, which might affect the production planning of automotive manufacturers;
3. Intensively emphasizing on new clients rather than existing clients, which might influence the dealers' sustainable sales capability.

3. Indicators Definition for Dealers' Performance Evaluation

To properly address the existing problems above-mentioned, the indicators definition is conducted by manufacturers. With regard to integrating sales and service processes in a supply chain, the balanced scorecard method is developed for dealers' performance evaluation.

3.1. *Balanced scorecard method*

In this paper, the balanced scorecard method, which was first coined by Kaplan and Norton (Kaplan and Norton, 1992) is used for performance management. It is related to a full-covering strategic evaluation indicator system, comprising both financial and non-financial indicators. More specifically, it connects the organizational performance evaluation with its long-term vision, mission statement and development strategy; it converts the organizational mission and tactics into tangible targets and assessment indicators, so as to link organizational plan with performance.

3.2. *Evaluation indicators definition*

By employing the theory of balanced scorecard method (Brewer and Speh, 2000; Ma, 2002; Shi and Cai, 2003), it is revealed that, while strategic objectives of manufacturers and dealers are defined, their development strategies should be further decomposed and analysed. In particular, this is aimed at promoting he profit, responding dealers' demands, and improving the sales and services capability. Based on this notion, the evaluation indicators are postulated from four aspects, namely financial issue, customer satisfaction, internal process, and research and innovation (see Fig. 3).

For the purpose of evaluation indicators definition, only the independency of dealer's operation is considered. In addition, it is found that the overall optimization on the supply chain should also be taken into account (Brewer and Speh, 2000; Yang *et al.*, 2008; Xiong *et al.*, 2006). In details, relevant indicators are provided as follows.

1. Financial issues. Due to the independent nature of dealer's operation, financial issue cannot be replied much on dealers' revenue and profit-related data. From the manufacturers' viewpoint, the major indicators

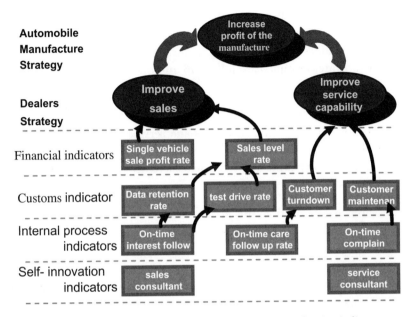

Fig. 3   Decomposition of supply chain strategies and evaluation indictors.

for financial issues are associated with the single vehicle sale profit and sales level rates during dealer's performance evaluation.

(a) The single vehicle sale profit rate is used to avoid vicious competitions among dealers in lowering the prices, so as to maintain the market stability.

(b) The sales level rate is used to ensure the dealers' fulfilment on the sales plan from manufacturers, so as to accurately make the production plan and to reduce the inventory of manufacturers.

2. Customer satisfaction. In essence, supply chain management is aimed to lower the cost of production and servicing by the means of information and resource sharing among the organizations and participants, so as to meet the increasingly-changing customer demands on high-quality products and services. The detailed indicators include the data retention, test drive, customer turndown and customer maintenance rates.

(a) The data retention and test drive rates are used to measure dealers' ability to attract new customers and to complete successful sales.

(b) The customer turndown and maintenance rates are used to measure dealers' ability to keep old customers and to maintain existing relationships.

3. Internal process. To meet the expectations of stakeholders and customers, the organization should implement a process that creates customer values. Corresponding indicators include on-time interest follow-up rate, on-time care follow-up and on-time complaint handling rates.

   (a) The on-time interest and care follow-up rates are used to improve the dealers' ability for proactive sales and services.
   (b) The on-time complaint handling rate is used to urge dealers to settle customer complaints and to resolve customer problems in time, so as keep old customers.

4. Research and innovation. It is mainly focused on the organization's long-term development. The primary indicators include sales and services consultant turnover rates.

The aforementioned four categories of indicators are interacted with each other to ensure the achievement of manufacturers' and dealers' strategic objectives.

4. Dealers' Performance Evaluation via ANP

4.1. *Analytic network process (ANP)*

The dealers' performance evaluation involves the operation of the entire automobile supply chain. Due to the indicators are inter-dependant with one another, an effective evaluation method should be investigated and employed to resolve this problem (Yan *et al.*, 2009a, b). Accordingly, the analytic network process (ANP) (Qi and Ding, 2006) is applied to analyze the inter-dependant indicators for final performance evaluation.

4.2. *Enabling factors of dealers' performance*

As above-mentioned, from a holistic viewpoint of manufacturers, the enabling factors of dealers' performance are related to sales capability and service capability (see Fig. 4). From the evaluation indicators determined by balanced scorecard method, four aspects of dealers' performance can be obtained, i.e. the financial condition, customer satisfaction, process efficiency and self-innovation. Each of them is composed of the sub-indicators, as listed in following table (Table 1).

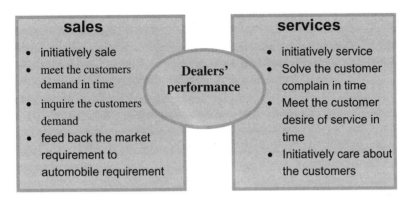

Fig. 4   Services process in the automotive supply chain.

Table 1   Sub-indicators of dealers' performance.

Sales	Service
Financial condition	Sales level rate
	Single vehicle sale profit rate
Customer satisfaction	Customer maintenance rate
	Customer turndown rate
	Test drive rate
	Data retention rate
Process efficiency	On-time complain handling rate
	On-time care follow up rate
	On-time interest follow up rate
Self innovation	Service consultant turnover rate
	Sales consultant turnover rate

4.3. *Procedure of dealers' performance evaluation*

The procedure of dealers' performance evaluation is presented as follows.

Step 1: Establish the judgement matrix. The judgement matrix is formed to compare the relative importance of lower-lever indicators against upper-lever indictors. In particular, the ANP adopts Satty ratings (Qi and Ding, 2006), as shown in Table 2.

1. Weights of two enabling factors. Compare the weights of sales and services capabilities (denoted by F, viz., $F = (f_1, \ f_2)$) (shown in Table 3).
2. Weight of the four aspects. The weights of four aspects (denoted by M, viz., $M = (m_1, \ m_2, \ m_3, \ m_4)$) (shown in Table 4).

Table 2   Definition of judgement matrix.

Scale	Definition
1	Same importance
3	Former is slightly more important than latter
5	Former is obviously more important than latter
7	Former is very much more important than latter
9	Former is extremely more important than latter
2,4,6,8	Middle value between above adjacent values
Reciprocal	Relative importance of latter against former

Table 3   Weights of two enabling factors.

	Sales capability	Service capability	Relatively weight
Sales capability (f_1)	1	1/2	0.3333
Service capability (f_2)	2	1	0.6667

Table 4   Weights of four aspects.

Enabling factor	Financial issue	Customer satisfaction	Process efficiency	Self-innovation	Relatively weight
Financial (m_1)	1	1/7	1/5	1/3	0.0666
Customer satisfaction (m_2)	7	1	2	2	0.4369
Process efficiency(m_3)	5	1/2	1	2	0.2979
Self-innovation (m_4)	3	1/2	1/2	1	0.1986

Step 2: Define the relative weights of indicators. Supposed the weights of indicators for each aspect be the financial condition, customer satisfaction, process efficiency and self-innovation in terms of the sales and services capabilities, respectively. They are represented as a_{ij}^k regarding two enabling factors and four aspects (where $k = 1, 2$; $i = 1, 2, 3, 4$; $j = 1, 2$). For example of sales factor, the weights of four indictors under customer satisfaction aspect could be represented as a_{21}^1, a_{22}^1, a_{23}^1, a_{24}^1, respectively. Similarly to service factor, the weights of the same indicators could be represented as a_{21}^2, a_{22}^2, a_{23}^2, a_{24}^2.

The relative weights of indicators for each enabling factor can also be calculated in that manner. For instance, Tables 5 and 6 list the customer satisfaction weights of indicators against sales and services enabling factors, respectively.

Table 5   Customer satisfaction weights of indicators against sales capability.

Sales capability	Data retention rate	Test drive rate	Customer turndown rate	Maintenance rate	Relatively weight
Data retention rate	1	3	5	9	0.4511
Test drive rate	1/3	1	7	9	0.4344
Customer turndown rate	1/5	1/7	1	1	0.0587
Maintenance rate	1/9	1/9	1	1	0.0557

Table 6   Customer satisfaction weights of indicators against services capability.

Sales capability	Data retention rate	Test drive rate	Customer turndown rate	Maintenance rate	Relatively weight
Data retention rate	1	1/2	2	1/6	0.1161
Test drive rate	2	1	1/4	1/6	0.1082
Customer turndown rate	4	4	1	1/2	0.3008
Maintenance rate	6	6	2	1	0.4749

Table 7   Definition of judgement matrix.

Aspect weight (m_i)	Indicator	Indicator weight (a_{ij}^k)	Adjusted indicator weight $(a_{ij}^{k'})$
0.0666	Single vehicle sale profit rate	0.2000	0.0133
0.0666	Sales level rate	0.8000	0.0533
0.4369	Data retention rate	0.4511	0.1971
0.4369	Test drive rate	0.4344	0.1898
0.4369	Customer turndown rate	0.0587	0.0256
0.4369	Maintenance rate	0.0557	0.0243
0.2979	On-time complain handling rate	0.1905	0.0567
0.2979	On-time care follow up rate	0.4762	0.1419
0.2979	On-time interest follow up rate	0.3333	0.0993
0.1986	Sales consultant turnover rate	0.6667	0.1324
0.1986	Service consultant turnover rate	0.3333	0.0662

Step 3: Calculate weights of indicators against each aspect. Upon completion of calculating the relative weights of each indicator, the weights of indicators are adjusted regarding different aspects. Supposed the adjusted weights of indicators be $a_{ij}^{k'}$, $a_{ij}^{k'} = a_{ij}^k . m_i$. The results are shown in Tables 7 and 8.

Table 8　Weights of indicators against services capability.

Aspect weight (m_i)	Indicator	Indicator weight (a_{ij}^k)	Adjusted indicator weight $(a_{ij}^{k'})$
0.0666	Single vehicle sale profit rate	0.5	0.0333
0.0666	Sales level rate	0.5	0.0333
0.4369	Data retention rate	0.1161	0.0507
0.4369	test drive rate	0.1082	0.0473
0.4369	Customer turndown rate	0.3008	0.1314
0.4369	maintenance rate	0.4749	0.2075
0.2979	On-time complain handling rate	0.5300	0.1579
0.2979	On-time care follow up rate	0.4038	0.1203
0.2979	On-time interest follow up rate	0.0662	0.0197
0.1986	sales consultant turnover rate	0.3333	0.0662
0.1986	service consultant turnover rate	0.6667	0.1324

Table 9　Synthetic weights of indicators.

Aspect weight	Indicator	Indicator weight	Sort
Financial issue	Single vehicle sale profit rate	0.0266	11
	Sales level rate	0.0400	10
Customer satisfaction	Data retention rate	0.0995	5
	Test drive rate	0.0948	7
	Customer turndown rate	0.0961	6
	Maintenance rate	0.1464	1
Process efficiency	On-time complain handling rate	0.1242	3
	On-time care follow up rate	0.1275	2
	On-time interest follow up rate	0.0462	9
Self-innovation	Sales consultant turnover rate	0.0883	8
	Service consultant turnover rate	0.1103	4

Step 4: Elaborate the impacts of sales and services capabilities on the enabling factors. By calculating the synthetic weights of different indicators, it is the sum of weights of all enabling factors multiply weights of adjusted indictors, viz., $w_i = a_{ij}^{1'} {}^* f_1 + a_{ij}^{2'} {}^* f_2$ (also see Table 9).

Step 5: Sort the evaluation indicators for dealers' performance. By applying the ANP technique, further analysis on indicators should be conducted using the future business areas for dealers, that is, improving the proactive follow-up post-sale services to ensure customers to choose dealers as their maintenance providers; quickly handling customer complaints and problems; promoting customer satisfactions; enhancing the post-sale service team; enabling the service team to provide customers with consistent and high-quality services; frequently communicating with potential customers by actively contacting with them and urging them on test drive so as to increase the deal rate.

4.4. *Method for dealers' performance evaluation*

The bespoke indictors are quantitatively calculated based on actual business data from dealers. Pertaining to the practical situations, the target values or the up-limits are defined for each indicator (referring to Table 10).

Table 10   Target rates of indicators.

Aspect weight	Indicator	Indicator weight	Assess target rate %	Definition
Financial issue	Single vehicle sale profit rate	0.0266	15	the higher the better, reach the up limit is full mark
	Sales level rate	0.0400	100	he higher the better, reach the up limit is full mark
Customer satisfaction	Data retention rate	0.0995	60	he higher the better, reach the up limit is full mark
	Test drive rate	0.0948	40	he higher the better, reach the up limit is full mark
	Customer turndown rate	0.0961	30	the lower the better, the higher rate , the lower mark
	Maintenance rate	0.1464	60	the higher the better, reach the up limit is full mark
Process efficiency	On-time complain handling rate	0.1242	100	the higher the better, reach the up limit is full mark
	On-time care follow up rate	0.1275	100	the higher the better, reach the up limit is full mark
	On-time interest follow up rate	0.0462	100	he higher the better, reach the up limit is full mark
Self-innovation	Sales consultant turnover rate	0.0883	40	the lower the better, the higher rate
	Service consultant turnover rate	0.1103	30	the lower the better, the higher rate

The full-mark for each indicator is 100. In case of higher marking, indictor score = indicator rate/assessment target rate * 100, whereas for lower marking, indicator score = (100% − indicator rate/assessment target rate) * 100. The final score for dealers' performance evaluation = $\sum$ (indicator score * indicator weight).

5. Case Study

By applying the proposed evaluation method, a case study was conducted based on three large-dimensional dealers for a specific automobile company. In order to promptly detect the dealers' operation problems and to consider the dealers' business turnover cycle, the evaluation period was set to be 1 month, December 2008 in particular. Table 11 shows the results from performance evaluation for the bespoke three dealers.

From the results obtained, it could be found that Dealers A and C performed well, whereas Dealer B lagged behind due to low scoring, especially in internal process efficiency and employee development. This indicated that Dealer B should improve such performances as pre-sale customer interaction, follow-up post-sale service, and team management.

Table 11 Results from performance evaluation for three dealers.

	Indicator	Dealer A	Dealer B	Dealer C
Financial issue	Single vehicle sale profit rate	2.13	2.02	2.29
	Sales level rate	4.00	3.80	3.80
Customer satisfaction	Data retention rate	7.96	7.46	7.96
	test drive rate	8.15	8.25	8.34
	Customer turndown rate	7.50	5.77	7.59
	maintenance rate	12.59	11.71	12.00
Process efficiency	On-time complain handling rate	11.80	10.56	11.67
	On-time care follow up rate	12.24	**7.65**	11.86
	On-time interest follow up rate	4.30	3.51	4.57
Self- innovation	Sales consultant turnover rate	7.06	**6.53**	7.95
	Service consultant turnover rate	9.82	8.82	10.15
The end performance evaluation result		87.54	76.09	88.19

6. Conclusions

In this study, a front market information integration model was established for manufacturers' and dealers' business. Based on this notion, the dealers' performance was evaluated from the automotive supply chain viewpoint. By using the balanced scorecard method, a breaking-down analysis was first conducted for the strategic objectives of an automotive supply chain. Subsequently, the evaluation indicators for dealers' performance were speculated in terms of four aspects, i.e., the financial condition, customer satisfaction, process efficiency and self-innovation. Next, the ANP technique was investigated and employed to obtain the indicator weights for dealers' performance evaluation. Consequently, the importance of indicators was sorted. To this end, the final results revealed that this approach assisted manufacturers to implement a more objective and effective assessment towards their dealers.

It could be found in this study that

1. compared with other dealers' performance evaluation methods, this approach was oriented from a manufacturer point of view;
2. this approach involved a more comprehensive method that concerned both short-term and long-term goal advancements of dealers; and
3. compared with traditional performance evaluation methods, such as time lag and short-term profit hunting, this approach focused intensively on the dealer' internal performance evaluation.

Acknowledgments

This work was supported by funding from Shanghai Science & Technology Committee (08ZR1409200), Shanghai Education Committee (09ZZ163, J50604), and Shanghai Maritime University (20080459).

References

1. Brewer, P. and Speh, T. (2000). *Journal of Business Logistics*, 75.
2. Cao, J.X., Shi, Q.X. and Lee, D.H. (2008). *Tsinghua Science & Technology*, 211.
3. Chen, C.H., Yan, W. and Chen, K. (2008). *International Journal of Computer Application in Technology*, 298.
4. Dreyer, D.E. (2000). *Supply Chain Management Review*, 30.
5. Kaplan, R. and Norton, D. (1992). *Harvard Business Review*, 71.
6. Ma, S.H., Li, H.Y. and Lin, Y. (2002). *Industrial Engineering and Management*, 5.

7. Ma, S.H. and Liu, X.G. (2007). *Automotive Engineering*, 174.
8. Qi, E.S. and Ding, X. (2006). *Journal of Harbin University of Commerce*, 11.
9. Shi, L.P. and Cai, X. (2003). *Journal of Harbin University of Commerce* (Social Science Edition), 84.
10. Xiong, S.G., Yi, S.P., Tang, P., Ping and Liu, F. (2006). *Journal of Chongqing University* (Natural Science Edition), 4.
11. Yan, W. Chen, C.H. Chang, D.F. and Chong, Y.T. (2009). *Advanced Engineering Informatics*, 201.
12. Yan, W. Chen, C.H. Huang, Y.F. and Mi, W.J. (2009). *Computers in Industry*, 21.
13. Yang, H.J., Sun, L.Y. and Gao, J. (2008). *Industrial Engineering and Management*, 96.
14. Zhang, Y. (2003). *Journal of Southwest University for Nationalities* (Natural Science Edition), 199.

PART II
PORTS AND LINERS OPERATIONS

CHAPTER 8

A YARD ALLOCATION STRATEGY FOR EXPORT CONTAINERS VIA SIMULATION AND OPTIMIZATION

Wei Yan, Junliang He and Daofang Chang

Logistics Engineering School, Shanghai Maritime University
1550 Pudong Avenue, Shanghai 200135, P.R. China
weiyan@shmtu.edu.cn

Container terminals, including shipping and land transportation, secure a crucial position in container transportation. In particular, container yard management, which involves diverse operational services, significantly affects the operational efficiency of the entire container terminal. Therefore, it is imperative to attain an efficient strategy to support the yard allocation for export containers. In this paper, a yard allocation model via objective programming was initially postulated based on a rolling-horizon strategy, which aims at allotting export containers into yard. Accordingly, the model's objective functions were subject to the minimum total distance of container transportation between storage blocks and berthing locations. This could balance the workloads amongst blocks. To resolve the NP-hard problem regarding the yard allocation model, a hybrid algorithm, which applies heuristic rules and genetic algorithm (GA), was then employed. Afterwards, a simulation model, which embeds the yard allocation model and algorithm, was developed to evaluate the proposed system. Subsequently, a case study was used for system illustration and simulation. Consequently, it could be found from computational results that this approach paves a venue for resolving the container yard allocation problem.

1. Introduction

As a hinge of global economy and trade, container terminals play an important role in a worldwide competition environment. It is well known that it becomes an imperative to improve operational efficiency of existing container terminals. The temporary storage of import and export containers, i.e., the storage space allocation problem, is critical to the container terminal services. The yard allocation for export containers at blocks affects directly on the moving distance of equipments and turnaround time of vessels. The container storing or retrieving process is associated with the time to arrange yard cranes, position and load/unload containers. As a container should be allotted to or picked up from somewhere in a block, other containers might be re-marshaled. In case when the export containers for the same vessel are placed intensively, it is likely more yard cranes are needed for this purpose. Thus, the workload balance between blocks is problematic. This results in a higher operating time and handling cost for yard cranes. Therefore, the rational yard allocation for export containers is imperative to the stowage planning and equipment scheduling.

Based on these understandings, a yard allocation model for export containers was developed based on a rolling-horizon strategy of objective programming. This is aimed at minimizing the total horizontal transportation distance and balancing the workloads among blocks. In order to resolve the NP-hard problem regarding the yard allocation problem, a hybrid algorithm, which integrates heuristic rules and genetic algorithm (GA), was employed. In this respect, the heuristic rules were postulated for generating feasible solutions, whereas the GA was applied for optimizing these solutions. Furthermore, a simulation model was deployed for system evaluation. Finally, a case study on a specific container storage yard was used for system illustration.

2. Related Work

Up to present, a number of researchers attempted to deal with the problems concerning the yard allocation problems for a container terminal. Current research work was focused on import containers, export containers, and a combination of import and export containers.

2.1. *Yard allocation for import containers*

Sgouridis *et al.* (2003) focused on the simulation of import containers that were transported by trucks. This approach was involved in a medium-size terminal in terms of an 'all-straddle-carrier' system. The simulated system was proposed for short- or medium-term planning using a process improvement strategy. Meanwhile, Bish (2003) proposed a strategy for determining a storage location for each import container, dispatching vehicles to containers, and scheduling the loading and unloading operations for cranes. This was aimed at minimizing the maximum time for vessel serving. Further studied, Wang (2007) postulated a plan-rolling model to resolve the mixed stocking for stochastic import containers in a container yard. To balance the travel and queuing time of containers, Cao *et al.* (2008) proposed a storage allocation strategy by integrating the integer programming, genetic algorithm and greedy heuristic algorithm. Similarly, Chang *et al.* (2008) synthesized the dynamic berth allocation and yard planning for import containers, which combined the heuristics algorithm and simulation optimization.

2.2. Yard allocation for export containers

To enhance the space and loading efficiency of export containers, Kim & Park (2003) developed a mixed integer programming model based on the sub-gradient optimization. Alternatively, Kim & Lee (2006) maximized the efficiency of yard trucks and transfer cranes. Kang & Ryu (2006) postulated a stacking strategy using the simulated annealing search. Furthermore, Zhang *et al.* (2007) proposed an optimization model of intra-bay relocation for export containers to minimize the total number of re-marshals and ensure the number of re-marshals within the average range. Meanwhile, Yan *et al.* (2008) combined yard crane scheduling and yard planning for export containers allocation. In this study, the hill-climbing algorithm, together with the best-first-search algorithm, was employed to resolve the NP-hard problem.

2.3. Combined yard allocation

Preston & Kozan (2001) minimized the turnaround time of container vessels. Zhang *et al.* (2003) proposed a yard allocation model for import and export containers to balance workloads amongst blocks. This study minimized the total transportation distance between yard and berth, which involved such terminal resources as quay and yard cranes, yard space and internal trucks. Based on these notions, Chen *et al.* (2004) narrowed the storage problem down and focused on the central allocation process to guarantee space efficiency. For the same purpose, Lim & Xu (2005) speculated a meta-heuristic procedure for yard allocation, i.e., critical-shaking neighborhood search. Moreover, Bazzazi *et al.* (2007) deployed a genetic algorithm (GA) strategy to resolve an extended storage space allocation problem (SSAP) simultaneously for import and export containers.

Similarly, Fu (2007) employed heuristic algorithms for space allocation through various time, quantity and container size.

Although intensive researches have been investigated into yard allocation, the yard allocation problem has not been well addressed for export containers rather than import containers. Compared with import containers, a more complicated problem-solving occurred pertaining to export containers. Therefore, it is imperative to develop a more effective yard allocation strategy for export containers. Thus, a comprehensive approach, along with multiple enabling objectives and constraints, is still lacking. As a result, a hybrid algorithm, which applies heuristic rules and genetic algorithm (DGA), was explored in this study.

3. Yard Allocation Modeling for Export Containers

3.1. *Problem description*

The turnaround time of vessels consists of loading and unloading time for containers. In order to reduce loading time, the storage locations for export containers should be selected for loading onto vessels efficiently. However, rational locations for export containers are cohesively associated with effective yard planning. In this paper, the number of containers for each vessel was determined for each block. To minimize turnaround time and handling cost of vessels, the workloads among blocks were balanced for each vessel, and the total distance of container transportation between the storage blocks and vessel berthing locations was minimized. In this regard, yard cranes in blocks were simultaneously served for vessels, and the berthing time of vessels was related to the maximal processing time of yard cranes. In general, the workload balancing on each block for vessels could reduce the completion time of vessels, and eliminate the traffic jam of equipments. In other words, the transportation distance between storage blocks and berthing locations affected directly on the turnaround time of vessels. If the transportation distance was shorter, the turnaround time would be less.

3.2. *Yard allocation modeling*

Owing to the uncertainty information of vessel arriving at port, a decision-making strategy, which synthetically considers all arriving vessels in 4 days, is developed using rolling-horizon approach (Zhang *et al.*, 2003). Meanwhile, a planning horizon of 4 days, each day divided into two 12-hour periods, is set. At the beginning of the first period, a storage space allocation plan is

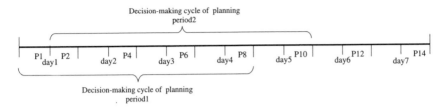

Fig. 1   Rolling-horizon strategy.

formed for the 8 periods within days 1–4. Only the plan of the first period is executed and a new 4-day plan is formed at the end of the first period. The details are shown in Fig. 1.

(i) Assumptions

The yard allocation model for export containers is developed based on the following assumptions:

1. The berth, berthing time and departing time of all arriving vessels in decision-making cycle are known;
2. It can be predicted according to statistics that the number, type and weight distribution of containers loading/unloading;
3. The number of quay crane scheduled for every vessel is estimated;
4. It accords with historical statistics that the number of export containers into yard, the retrieved number of containers and the retrieved time.

(ii) Notation

TP	The total number of planning periods in a decision-making cycle.
$TP = 8$	Planning period is denoted t. Only the plan of the first period is executed;
NA	The total number of blocks in the yard;
P	The currently decision-making cycle;
VP_t	The set of all vessels needing yard planning in decision-making cycle P;
VP_{jt}	Vessel j in period t;
NVP_{ti}	The set of vessels which have been allocated in block i before period t;
B_{jt}	The berthing place of vessel VP_{jt};
d_{ij}	Distance between block i and the berthing place of vessel VP_{jt} in period t;

$N2_{jt}$	The number of 20-foot export containers of vessel VP_{jt} in period t;
$N4_{jt}$	The number of 40-foot export containers of vessel VP_{jt} in period t;
R_i	The lanes of block i;
T_i	The tiers of block i;
OPL_j	The estimated number of quay crane scheduled for vessel VP_{jt} in period $t\,K_{jt}$. Up to period t, the periods that export containers of vessel VP_{jt} having arrived at container terminal;
STH_{it}	The set of start times for all vessels in block i in period t. $STH_{it} = \{STH_{it1}, STH_{it2}, \ldots, STH_{itn}\}$;
ETH_{it}	The set of end times for all vessels in block i in period t. $ETH_{it} = \{ETH_{it1}, ETH_{it2}, \ldots, ETH_{itn}\}$;
ST_{tj}	The loading start time of vessel VP_{jt} in period t;
ET_{tj}	The loading end time of vessel VP_{jt} in period t;
$N2_{ijtk}$	The total number of 20-foot export containers allocated in block i that arrive at the container terminal in period $t-k$;
$N4_{ijtk}$	The total number of 20-foot export containers allocated in block i that arrive at the container terminal in period $t-k$;
$NU_{i(t-1)}$	The number of empty bay at the end of period $t-1$;
$NUB_{i(t-1)}$	The number of empty 40-foot bay (Two adjacent bays) at the end of period $t-1$;
λ	Expansion coefficient.

(iii) Decision Variables

$$A_{ijt} = \begin{cases} 1, & \text{Block } i \text{ is allocated for vessel } VP_{jt} \text{ in period } t \\ 0, & \text{Block } i \text{ is not allocated for vessel } VP_{jt} \text{ in period } t \end{cases}$$

$$H_{ijt_k} = \begin{cases} 1, & \text{Block } i \text{ is allocated for vessel } VP_{jt} \text{ in period } t-k \\ 0, & \text{Block } i \text{ is not allocated for vessel } VP_{jt} \text{ in period } t-k \end{cases}$$

$N2_{ijt}$—The total number of 20-foot export containers allocated in block i that arriving at the container terminal in period t;

$N4_{ijt}$—The total number of 20-foot export containers allocated in block i that arriving at the container terminal in period t.

(iv) Mathematical Models

The decision-making objectives are presented as follows.

1. Minimizing the total distance of all vessels to transport the containers between their storage blocks and the vessel berthing locations;
2. Balancing the workload among blocks allocated vessel VP_{jt} in period t;
3. Balancing the workload among all blocks.

The Mathematical models are presented as follows.

$$f_1 = \text{Min} \sum_{t=1}^{TP} \sum_{j \in VP_t} \sum_{i=1}^{NA} (N2_{ijt} + N4_{ijt}) \cdot A_{ijt} \cdot d_{ijt}, \tag{1}$$

Equation 1 is the first objective to minimize the total distance of all vessels to transport the containers between their storage blocks and the vessel berthing locations, which synthetically considers all arriving vessels in 4 days.

$$f_2 = \text{Min} \left\{ \underset{\{i\}}{\text{Max}} \left[\sum_{k=1}^{K_{jt}} (N2_{ijtk} + 2 \cdot N4_{ijtk}) \cdot H_{ijt_k} \right. \right.$$

$$+ \left. \sum_{t=1}^{8-K_{jt}} (N2_{ijt} + 2 \cdot N4_{ijt}) \cdot A_{ijt} \right]$$

$$- \underset{\{i\}}{\text{Min}} \left[\sum_{k=1}^{K_{jt}} (N2_{ijtk} + 2 \cdot N4_{ijtk}) \cdot H_{ijt_k} \right.$$

$$+ \left. \left. \sum_{k=1}^{K_{jt}} (N2_{ijtk} + 2 \cdot N4_{ijtk}) \cdot H_{ijt_k} \right] \right\}, \tag{2}$$

Equation 2 is the second objective to balance the workload among blocks allocated for vessel VP_{jt}, which minimizes the margin between the block with the maximum export containers of VP_{jt} and the block with the minimum export containers of VP_{jt}.

$$f_3 = \text{Min} \left\{ \sum_{t=1}^{TP} \underset{\{i\}}{\text{Max}} \left[\sum_{j \in VP_t} (N2_{ijt} + 2 \cdot N4_{ijt}) \cdot A_{ijt} \right] \right.$$

$$\left. - \underset{\{i\}}{\text{Min}} \left[\sum_{j \in VP_t} (N2_{ijt} + 2 \cdot N4_{ijt}) \cdot A_{ijt} \right] \right\}, \tag{3}$$

Equation 3 is the third objective to balance the workload among all blocks, which minimizes the margin between the block with the maximum workloads and the block with the minimum workloads. This is used to avoid the traffic jam.

$$\text{Min}\{\omega_1^* f_1, \ \omega_2^* f_2, \ \omega_3^* f_3\}, \tag{4}$$

Equation 4 is a multi-objective function formed via Equations 1, 2 and 3. $\omega 1$, $\omega 2$ and $\omega 3$ are respectively the weights of Equations 1, 2 and 3.

$$t \in TP, \quad j \in VP_t, \tag{5}$$

Equation 5 is a constraint to ensure that the planning period is in the decision-making cycle and the vessel is needed for yard planning in period t.

$$N2_{jt} = \sum_{i=1}^{NA} A_{ijt}^* N2_{ijt}, \tag{6}$$

Equation 6 is a constraint to ensure that the total number of 20-foot export containers allocated in blocks i is the sum of these containers assigned to all the blocks in period t.

$$N4_{jt} = \sum_{i=1}^{NA} A_{ijt}^* N4_{ijt}, \tag{7}$$

Equation 7 is a constraint ensure that the total number of 40-foot export containers allocated in blocks i is the sum of these containers assigned to all the blocks in period t.

$$\sum_{k=1}^{K_{jt}} \sum_{i=1}^{NA} H_{ijt_k} + \sum_{t=1}^{8-K_{jt}} \sum_{i=1}^{NA} A_{ijt} = 2 \cdot OPL_{jt}, \tag{8}$$

Equation 8 is a constraint to ensure that the total number of blocks allocated to any vessel is two times that of quay cranes scheduled.

$$\forall\, t = (1, 2, \ldots, 8),$$

$$\lambda \cdot \left(\sum_{j \in VP_t} N2_{jt} + 2 \cdot \sum_{j \in VP_t} N4_{jt} \right) \qquad (9)$$

$$\leq \sum_{i=1}^{NA} A_{ijt} \cdot NU_{i(t-1)} \cdot [R_i \cdot T_i - (T_i - 1)],$$

Equation 9 is a constraint to ensure that the allowable stacks of all blocks will not be less than the total number of export containers of all vessels in period t. If $\exists\, j \in VP_t$ and $AL_{ijt} = 1$, the constraint can be ensured. The constraint is only used for the 20-foot export containers. As well, a special constraint for the 40-foot export containers should be given, because the 40-foot containers should be stored in two adjacent bays.

$$\forall\, t = (1, 2, \ldots, 8), \lambda \cdot \sum_{j \in VP_t} N4_{jt}$$

$$\leq \sum_{i=1}^{NA} A_{ijt} \cdot NUB_{i(t-1)} \cdot [R_i \cdot T_i - (T_i - 1)], \qquad (10)$$

Equation 10 is a constraint to ensure that the allowable stacks of all blocks will not be less than the total number of 40-foot export containers of all vessels in period t.

$$\forall\, m \in NVP_{ti}, A_{ijt} \cdot [STH_{mt} - ET_{jt}] \cdot [ETH_{mt} - ST_{jt}] > 0. \qquad (11)$$

Equation 11 is a constraint to ensure that the handling time of all vessels, for which containers stored in the same block, will not be overlapped.

4. Yard Allocation Algorithm for Export Containers

Specifically for this study, the heuristic algorithm is used for reducing feasible solution scale via constraints. Meanwhile, the genetic algorithm is used for optimizing feasible solution to gain the approximate optimal solution for yard allocation for export containers.

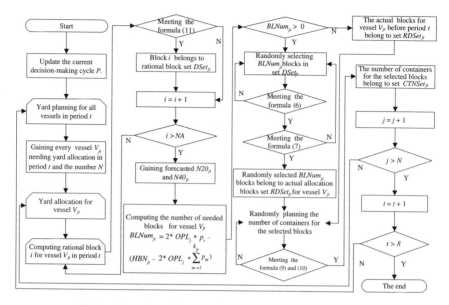

Fig. 2   Procedure of heuristic algorithm for feasible solution.

4.1. Heuristic algorithm for feasible solution

The heuristic algorithm, which generates feasible solution for yard allocation for export containers, is shown in Fig. 2.

The heuristic algorithm is involved in five rules provided as follows.

Rule 1: $A_{ijt} \cdot [STH_{mt} - ET_{jt}] \cdot [ETH_{mt} - ST_{jt}] > 0$. This rule ensured that the handling time of all vessels, from which containers are stored in the same block, will not overlap.

Rule 2: $N2_{jt} = \sum_{i=1}^{NA} A_{ijt} \cdot N2_{ijt}$. This rule ensured that the total number of 20-foot export containers allocated in block i is the sum of these containers assigned to all the blocks in period t.

Rule 3: $N4_{jt} = \sum_{i=1}^{NA} A_{ijt} \cdot N4_{ijt}$. This rule ensured that the total number of 40-foot export containers allocated in block i is the sum of these containers assigned to all the blocks in period t.

Rule 4: $\lambda \cdot (\sum_{j \in VP_t} N2_{jt} + 2 \cdot \sum_{j \in VP_t} N4_{jt}) \le \sum_{i=1}^{NA} A_{ijt} \cdot NU_{i(t-1)} \cdot [R_i \cdot T_i - (T_i - 1)]$. This rule ensured that the available stacks of all blocks are not fewer than the total export containers of all vessels in period t.

Rule 5: $\lambda \cdot \sum_{j \in VP_t} N4_{jt} \leq \sum_{i=1}^{NA} A_{ijt} \cdot NUB_{i(t-1)} \cdot [R_i \cdot T_i - (T_i - 1)]$. This rule ensured that the available stacks of all blocks are not fewer than the total 40-foot export containers of all vessels in period t.

4.2. Procedure of genetic algorithm

The procedure of genetic algorithm is presented as follows.

Step 1: Encoding representation. The different planning periods in decision-making cycle represent chromosomes. The chromosome consists of four dimensions for the indices regarding vessel, allocated block, number of 20-foot export containers and number of 40-foot export containers. The vessel gene is arranged in time sequence arrived at terminal, i.e., $V1, V2, \ldots, Vn$. The allocated block gene consists of two numbers, which are at odd bit and right adjacent even bit, respectively. The genes of number of 20-foot export containers and number of 40-foot export containers are composed like the allocated block gene.

Step 2: Population initialization. The first generation population of every processor is randomly generated with heuristic algorithm (Fig. 2). The population size n and the number of processors m are set. The sub-population size of every processor is n/m.

Step 3: Judging individual feasibility (Fig. 2). If the individual is feasible, it will be held; otherwise, it will be mutated.

Step 4: Fitness evaluation. Supposed that Equations 1, 2 and 3 are unitary.

(i) If Equation 1 is unitary, the minimum and the maximum distance to transport the containers between their storage blocks and the vessel berthing locations are set as $l_{\min}$ and $l_{\max}$, respectively. The unitary formula is presented as follows.

$$f_1' = \frac{f_1 - \sum_{j \in VP_t}(N2_{jt} + N4_{jt}) \cdot l_{\min}}{\sum_{j \in VP_t}(N2_{jt} + N4_{jt}) \cdot l_{\max} - \sum_{j \in VP_t}(N2_{jt} + N4_{jt}) \cdot l_{\min}},$$
$$(12)$$

(ii) If Equation 2 is unitary, the minimum and the maximum imbalance between the block allocated for some vessel are set as $CTN_{\mathrm{Min}j}$ and $CTN_{\mathrm{Max}j}$, respectively. The unitary formula is presented as follows.

$$f_2' = \sum_{j \in VP_t} \frac{f_{2j} - CTN_{\mathrm{Min}j}}{CTN_{\mathrm{Max}j} - CTN_{\mathrm{Min}j}}, \qquad (13)$$

(iii) If Equation 3 is unitary, the minimum and the maximum imbalance among all blocks are set as $ALLCTN_{\min}$ and $ALLCTN_{\max}$, respectively. The unitary formula is presented as follows.

$$f_3' = \frac{f_3 - ALLCTN_{\min}}{ALLCTN_{\max} - ALLCTN_{\min}}, \tag{14}$$

The unitary objective function is defined as $f = \text{Min}(\omega_1^* f_1' + \omega_2^* f_2' + \omega_3^* f_3')$. The fitness function of the genetic algorithm is defined as $f(k) = 1/(\omega_1^* f_1' + \omega_2^* f_2' + \omega_3^* f_3')$.

Step 5: Selection strategy. The individual is selected in the mating pool of parents with 'roulette wheel' sampling. The process is presented as follows.

(i) Calculate the sum of fitness of all individuals in sub-population (Equation 15). Notation n is the sub-population size.

$$SUM = \sum_{i=1}^{n} F(x^i), \tag{15}$$

(ii) Calculate the selected probability of individual (Equation 16).

$$P(x^j) = \frac{F(x^j)}{\sum_{i=1}^{n} F(x^i)}, \quad (j = 1, 2, \ldots, n). \tag{16}$$

(iii) Calculate the cumulative probability of individual.
(iv) Generate a random number RM between 0 and 1.
(v) If the RM is between the cumulative probability of two individuals, the individual, which cumulative probability is greater, will be selected.

Step 6: Selection, crossover and mutation are implemented with 'steady state' method. The single-point crossover, which crosses the entire gene including block, is used for the number of 20-foot export containers and number of 40-foot export containers. In this paper, a mutation operator, called gauss mutation, is also used. The number of substitutes is decided by designating the percentage of replacement to replace each generation. The selected individual will be brought in the pool of parents for the next generation.

Step 7: Migration operation. In this paper, the migration operation is conducted according to the migration number, migration frequency and topology relation among all populations. The migration frequency and migration number are set as 3 and 5, respectively. The topology relation among all populations is set as circularity. The migration individual will be brought in the pool of parents for the next generation. If the migration operation is done, every sub-population evolves respectively, and then goes to Step 3; otherwise, directly go to Step 3. If the stoppage rules (the maximum elapsed generation) are met, the algorithm stops.

Step 8: Stoppage rule. The algorithm is terminated regarding the experimentally-determined maximum elapsed generation.

5. Simulation Model

To evaluate the proposed strategy on yard allocation for export containers, an effective method is imperative. Thus, it is quite complicated to apply such mathematical models as operational research. Accordingly, simulation is an efficient tool to evaluate the performance of complex systems. Hence, the simulation technology could only evaluate an existing approach rather than an optimized strategy. Therefore, a combination of simulation and optimization technologies was proposed in this study, i.e., the yard allocation model and algorithm.

5.1. *Simulation framework*

A simulation model was established using eM-Plant^{TM} (Fig. 3). As illustrated in Fig. 4, this model comprised five interacting modules, namely

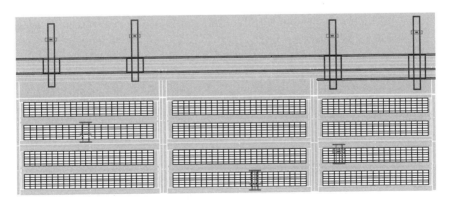

Fig. 3   Illustration of yard allocation simulation.

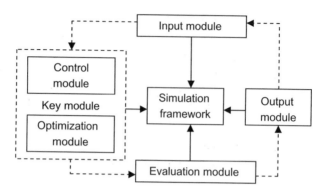

Fig. 4   Framework of simulation strategy.

input, control, evaluation, optimization and statistical output modules. In details, the yard allocation model for export containers was embedded in the control module, whereas the hybrid algorithm was contained in the optimization module. Subsequently, the feasible solutions, together with constraints, were generated via the heuristic algorithm. Consequently, the feasible solutions were optimized using the simulation-embedded DGA algorithm. In addition, the evaluation indices were subtracted for the evaluation module.

5.2. *Input parameters*

The following parameters should be initialized prior to simulation.

1. The parameters of container terminal layout included the quayside length, water depth, number of blocks, number of bays, lanes and tiers of blocks, coordinates of blocks, coordinates and length of tracks, coordinates of gateways, etc.
2. The parameters of mechanism included the size and velocity parameters.
3. The statistical distribution included the time interval distribution of arrived vessels, the time interval distribution of external trucks, distribution of vessel types, distribution of container types, distribution of containers, etc.

5.3. *Simulation process*

The simulation process of yard allocation for export containers was presented in Fig. 5.

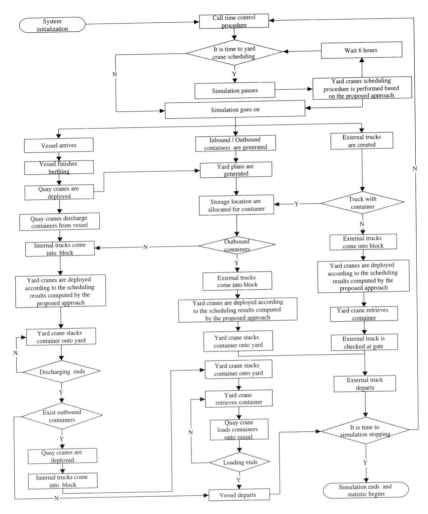

Fig. 5   Simulation process of yard allocation.

5.4. *Statistical simulation indices*

The following statistical indices should also be obtained during simulation.

1. The loading time of export containers onto vessels.
2. The total horizontal transportation distance between storage blocks and berthing locations for loading.
3. The balance among blocks for export containers, i.e., the gap between blocks with the maximum and minimum export containers.

6. Case Study

A case study on a specific container terminal was conducted. In details, a continuous 4-day data was stochastically selected (Table 1) to study the yard allocation for export containers. In particular, the container terminal possesses 32 blocks, together with 35 bays, 5 tiers and 7 lanes, handled by RTGCs, where each bay possesses a storage capacity of 31 containers. Meanwhile, there are 4 berths with quayside length of 1,100 meters.

The yard allocation for export containers was optimized via the simulation model. The comparison between optimized results and actual results was listed in Table 2, while the optimization process was shown in Fig. 5. In this case, some indices, i.e., the population size, sub-population size, crossover probability, Gaussian mutation probability, maximum elapsed generation and percentage of replacement, were initially set as 100, 50, 0.8, 0.05, 40 and 20%, respectively.

As shown in Table 2, the total vessel loading time of 13 vessels improved 8.10% against the actual operation, the total horizontal transportation distance of 13 vessels improved 9.13% and imbalance among all blocks improved 35.61%.

The computational results from yard allocation for export containers were discussed as follows. The allocated blocks of export containers for vessels were shown in Table 1. Meanwhile, the computational time of hybrid algorithm was approximately taken for 21 minutes.

Table 1   Data of arrived vessels in four consecutive days.

Vessel name	TEU		Number of load port	Number of weight	Berth of vessel
	20-foot.	40-foot.			
v01	283	121	9	4	3
v02	639	274	7	3	4
v03	250	107	6	4	2
v04	112	48	3	2	1
v05	289	124	4	1	3
v06	188	80	5	1	1
v07	255	109	10	5	1
v08	497	213	8	2	2
v09	524	224	8	3	3
v10	1112	476	12	5	3
v11	622	266	11	5	4
v12	522	223	7	4	3
v13	491	210	5	3	4

Table 2   Comparison between optimized and actual results.

Vessel name	Vessel loading (h)		Horizontal transportation distance (km)		Imbalance among blocks	
	Optimized	Actual	Optimized	Actual	Optimized	Actual
v01	3.9	4.3	267.8	294.0	3	5
v02	6.8	7.3	547.3	586.1	5	8
v03	4.0	4.1	250.4	264.9	5	8
v04	3.1	3.3	105.4	116.4	5	7
v05	4.3	4.7	254.4	288.4	5	7
v06	4.4	4.7	151.1	168.6	6	9
v07	4.1	4.6	318.8	364.2	5	8
v08	5.5	5.9	489.9	543.9	6	8
v09	4.9	5.2	573.6	613.7	4	7
v10	7.7	8.5	1252.9	1421.5	6	10
v11	5.4	6.0	665.9	720.8	6	10
v12	5.1	5.6	599.7	653.1	5	7
v13	4.2	4.5	575.5	625.0	5	7

To synthetically evaluate computational results via simulation, comparisons were conducted based on the results between the actual, manual and proposed approaches (shown in Table 2). In details,

1. Against the actual operation, the total loading time onto vessels, the total horizontal transportation distance for 13 vessels and the imbalance among blocks was improved by 8.10%, 13.37% and 36.19%, respectively.

2. Through the simulation on one month's data of the container terminal, it could be found from the experimental results that the average time of vessels staying at port reduces 9.2%.

To verify the effectiveness and reliability of the proposed yard allocation approach for export containers, the numerical experiments with regard to diverse yard occupation ratios were completed based on the data from Table 1. In this regard, the yard occupation ratios were predetermined as 40%, 60% and 80% for three experiments, respectively. Pertaining to statistical indices, the results were then compared with those from actual approach.

Tables 3 listed the comparisons of experiments regarding the yard occupation ratios of 40%, 60% and 80%, respectively. From the results in Table 3, all indices of bespoke three experiments obtained from the proposed approach were better than those obtained from actual approach. Furthermore, when the yard occupation ratio was lower, the improvement

Table 3   Comparisons among three experiments using yard occupation ratio.

	Yard occupation ratio		
	40%	60%	80%
Statistical indices	Improvement	Improvement	Improvement
Vessel loading (h)	8.10	16.53	23.58
Horizontal transportation distance (km)	9.13	16.88	26.20
Imbalance among blocks	35.61	37.38	36.06

of total loading time onto vessels and the total horizontal transportation distance were higher. When the yard occupation ratio tended to be higher, which was resulted from fewer blocks available, the improvement of imbalance among blocks might be reduced.

7. Conclusions

The efficiency of container terminals is significantly dependent on the resource allocation at diverse operational stages. Hence, the resource of container yard is costly and can be the bottleneck in container handling process. Accordingly, a yard allocation model was established for export containers based on a rolling-horizon strategy. This was aimed at minimizing the total horizontal transportation distance and balancing the workloads among all blocks.

In order to resolve the NP-hard problem regarding the yard allocation problem, a hybrid algorithm, which integrates heuristic rules and GA, was investigated. In details, the heuristic rules were developed for generating feasible solutions, while the GA was applied for optimizing these solutions. Eventually, a case study on a specific container storage yard was used for system illustration.

Hence, there still existed a space for improvement, e.g., the impacts on the berth allocation were not well addressed in this approach. As such, the possible extensions could be adapted to this study as future directions.

Acknowledgments

This work was supported by funding from Shanghai Science & Technology Committee (08ZR1409200), Shanghai Education Committee (09ZZ163, J50604), and Shanghai Maritime University (20080459, 20100130).

References

1. Bazzazi, M., Safaei, N. and Javadian, N. (2007). *Computers and Industrial Engineering*, In Press.
2. Bish, E.K. (2003). *European Journal of Operational Research*, 83.
3. Cao, J.X., Shi, Q.X. and Lee, D.H. (2008). *Tsinghua Science & Technology*, 211.
4. Chang, D.F., Yan, W., Chen, C.H. and Jiang, Z.H. (2008). *International Journal of Computer Applications in Technology*, 272.
5. Cheng, P., Fu, Z., Lim, A. and Rodriques, B. (2004). *IEEE Transactions on Automation Science and Engineering*, 26.
6. Fu, Z., Li, Y., Lim, A. and Rodriques, B. (2007). *Journal of the Operational Research Society*, 797.
7. Kang, J. and Ryu, K.R. (2006). *Journal of Intelligent Manufacturing*, 399.
8. Kim, H.K. and Lee, J.S. (2006). *International Conference on Computational Science and Its Applications*, 564.
9. Kim, K.H. and Park, K.T. (2003). *European Journal of Operational Research*, 92.
10. Lim and Xu, Z. (2006). *European Journal of Operational Research*, 1247.
11. Preston, P. and Kozan, E. (2001). *Computers and Operations Research*, 983.
12. Sgouridis, S.P., Makris, D. and Anqelides, D.O. (2003). *Journal of Waterway, Port, Coastal and Ocean Engineering*, 178.
13. Wang, B. (2007). *System Engineering Theory and Practice*, 147.
14. Yan, W., Huang, Y.F. and He, J.L. (2008). *International Journal of Computer Applications in Technology*, 254.
15. Zhang, Y.W., Mi, W.J., Chang, D.F., Yan, W. and Shi, L.D. (2007). *IEEE International Conference on Automation and Logistics*, 776.

CHAPTER 9

INTEGRATION OF AGVS IN INTERMODAL RAIL OPERATIONS AT DEEP SEA TERMINALS

Bernd H. Kortschak

Business Administration & Logistics,
Faculty Business, Logistics, Transport
University of Applied Sciences Erfurt, Germany
kortschak@fh-erfurt.de

Rail links in sea-terminals are often separated by several means: fences, barriers, special gates, costly haulage to overcome the interfaces. Some earlier solutions are dealt with — like to put a quai-track underneath the gantry crane to get direct transfer of containers. It will be shown in the paper that this solution is not state of the art for various reasons, the most important is that the gantry crane loading/unloading a 10,000 TEU vessel is not prepared to move forward and backward just to put the right container on the right wagon. The solution has to be found by including shuttle trains in the AGV regime of the terminal for better productivity and sufficient flexibility. The paper will outline which criteria to be dealt with to achieve that goal of better economic performance.

1. Introduction

Deep sea terminals face a multiplicity of criteria on port operations, e.g., short turnaround time, short dwelling time of containers in the port, enough stacking areas for peak demand. Different concepts have been deployed, even spatial differentiation between the deep sea transshipment and hinterland connections[x] in the Agile Port System in the U.S.[y] In this paper, the focus lies on the integration of all functions in a deep sea terminal, i.e., to link all modes within the terminal area; to integrate the functions despite of spatial constraints. Until now railway links to deep sea terminals are operated separately, which incurs additional costs for transfer and thus hinders competition strength of rail versus road.

To cater for growing container volumes, bigger ships demand for a strong hinterland connection provided by mass volume transport modes, such as inland waterways or rail. Whereas inland waterways depend on natural geographical landscape, rail services can be deployed almost everywhere in hinterland connections given that the volume of containers can justify the construction cost. Therefore the aim of this paper is to answer the question on how rail services can be integrated in terminal operations so as to increase their operation efficiency as illustrated by Fig. 1.

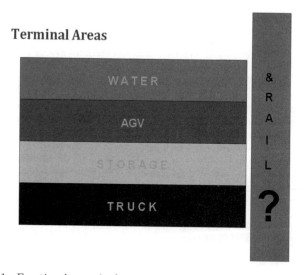

Fig. 1   Functional areas in deep sea terminals without rail integration.

---

[x]Some intermodal transfers take place at one point, while others involve two or more locations. The former is much more efficient. Muller, G.: Intermodal Freight Transportation, 3rd ed., Landsdowne VA 1995, p. 144.
[y]Vickermann, M.J. (1999). Agile Port Concept, http://www.transystems.com.

2. Earlier Attempts to Address the Problem

2.1. *Fixed rail mounted gantry cranes linking ship to shore crane with stacking area and hinterland modes rail and road*

This approach has been investigated by Howaldtswerke Deutsche Werft AG[z] for the Terminal of the Future which can serve vessels with size of up to 8000 TEU. It is implemented by overlapping yard crane services in a stacking corridor to enable the linkages between different transport modes.

2.2. *NOELL — an approach by K.-P. FRANKE*

The main idea behind this logistical concept is to load and unload large vessels on a reduced area of land with minimum impact on the inland public traffic system and the environment. In addition,[aa] the system is targeted on

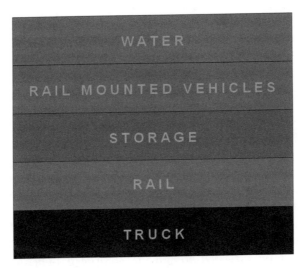

Fig. 2   The integration of rail with rail mounted vehicle.

---

[z]Howaldtswerke — Deutsche Werft, AG *et al.* (1997). Container-Transport-systeme, der Zukunft, Projekt, gefördert vom Bundesminister für Bildung, Wissenschaft, Forschung und Technologie, Förderkennzeichen 18S0071 A, Kiel.
[aa]Franke, K.-P. (2008). A technical approach tot he Agile Port System, in: Koenings, R., Priemus, H. and Nijkamp, P. (eds.): the Future of Intermodal Freight Transport Operations, Design and Policy, p. 137.

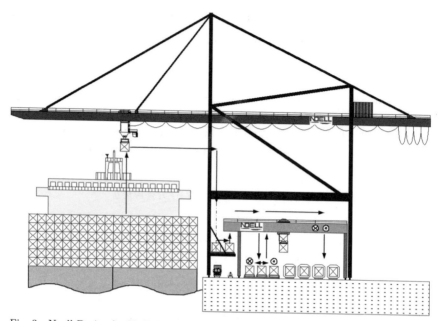

Fig. 3   Noell Design by K.-P. Franke of the Efficient Marine Terminal: Direct handling of containers between vessel and trains.[cc]

maximizing port productivity by transshipping boxes directly from vessels to trains and vice versa at the quay (shown in Fig. 3).[bb]

It incorporates the following features:

- Trolleys and ship-to-shore cranes are able to unload containers to a platform in the quayside portal, where the twist locks from deck containers can be removed
- A conveyor to move containers from the lashing position on the platform to a second position underneath a rail mounted gantry (RMG) cantilever, which could be extended to provide additional buffer-space. It was realized years ago by Matson Terminal, Los Angeles.

[bb]Franke, K.-P. (2008). A technical approach tot he Agile Port System, in: Koenings, R., Priemus, H. and Nijkamp, P. (eds.): the Future of Intermodal Freight Transport Operations, Design and Policy, p. 137.
[cc]Franke, K.-P. (2008). A technical approach tot he Agile Port System, in: Koenings, R., Priemus, H. and Nijkamp, P. (eds.): the Future of Intermodal Freight Transport Operations, Design and Policy, p. 139.

- RMGs operate under the portal of the ship-to-shore cranes; and they could cover four rail lanes and a three-lane wide box mover.
- There are two extra lanes under the lashing platform for the ship-to-shore cranes.

The big advantage of this new design concept is that yard transfer vehicles are not required, which could save a great deal of machinery and labor. When serving the vessel, one duty of the RMG would be to take containers from the platforms and place them on the linear-motor-based transfer system or the rail cars on the shortest possible way. The linear motor lanes could serve additional RMG loading and unloading along the trains as well as a buffer stack. The linear motor system would allow boxes being out of sequence to be held aside and shuffled without interrupting the ship-to-shore import — export cycle. Five to eight RMGs could serve five ship-to-shore cranes between them.[dd]

By considering other functions of a terminal into account, the spatial scheme is given by Fig. 4.

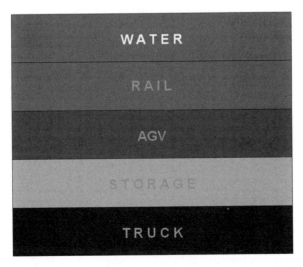

Fig. 4   The functional scheme of the NOELL solution by K.-P. Franke.

[dd]Franke, K.-P. (2008). A technical approach tot he Agile Port System, in: Koenings, R., Priemus, H. and Nijkamp, P. (eds.): the Future of Intermodal Freight Transport Operations, Design and Policy, p. 138.

3. The AGV-solution to Integrate Railway Operations in Deep Sea Terminals

In 2006, Duinkerken *et al.* compared multi-trailer-systems (MTS), automated-guided-vehicles (AGVs) and automated-lifting-vehicles (ALVs).[ee] Although ALVs are found to be superior, because they do not have to wait for cranes to load or unload,[ff] they are not considered here, because they need more space to be operated. The multi-trailer system uses manned trucks pulling trains of five trailers.[gg] 20 manned trucks and 130 sets of trains of five trailers have to be operated at least at Maasvlakte[hh] compared to Hamburg's Container Terminal Altenwerder (CTA) where a single trailer system is operated by using 12 traction units and about 300 trailers. CTA uses less resources due to the fact that the distance to be covered at Maasvlakte amounts up to 6 km whereas the rail link at CTA is only less than 1 km since ITT is close to the stacking area — less than 1 km distance has to be covered — even if the transfer to the train takes place at the opposite site of the rail terminal. See Fig. 5 below:

Operating an AGV system directly with the rail cranes could enhance crane productivity. The longitudinal movements of the crane for correct positioning of the containers are avoided; thus the loading and unloading activities can take place with a minimum of transfer time. Furthermore, the direct access from the rails to the ships and shore cranes enables faster transshipment times and reduces peak loads at the stacking areas. Hence, it increases the competitiveness of rails against roads because the duration for transfer between rails and deep sea ships may be accelerated. The only difficulty is to find a reserved space for AGVs' movements to and from the rail link terminal. If the rail terminal is parallel to the shore besides the stacking area, one solution might be that AGVs pass by the stacking area, the dead end of truck, rail terminal area, and move parallel to the

---

[ee] Duinkerken, M.B., Deekker, R., Kurstjens, S.T.G.L., Ottjes, J.A. and Dellaert, N.P. (2006). Comparing transportation systems for inter-terminal transport at the Maasvlakte container terminals, in: OR Spectrum 28, p. 469.

[ff] Duinkerken, M.B., Deekker, R., Kurstjens, S.T.G.L., Ottjes, J.A. and Dellaert, N.P. (2006). Comparing transportation systems for inter-terminal transport at the Maasvlakte container terminals, in: OR Spectrum 28, p. 471.

[gg] Duinkerken, M.B., Deekker, R., Kurstjens, S.T.G.L., Ottjes, J.A. and Dellaert, N.P. (2006). Comparing transportation systems for inter-terminal transport at the Maasvlakte container terminals, in: OR Spectrum 28, p. 470.

[hh] Duinkerken, M.B., Deekker, R., Kurstjens, S.T.G.L., Ottjes, J.A. and Dellaert, N.P. (2006). Comparing transportation systems for inter-terminal transport at the Maasvlakte container terminals, in: OR Spectrum 28, p. 485.

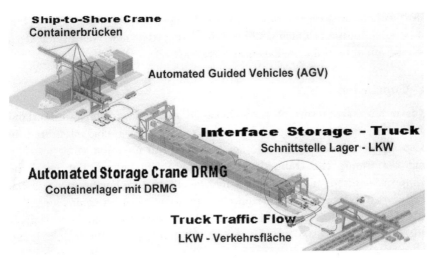

Fig. 5 Present scheme of operations at container terminal altenwerder source: Engelhardt, T. and Müller-Elsner, H. (2003). Ein Gigant mit Gehirn, in: GEO, 11, pp. 114–140, pp. 132–133.

The integration of AGV at the „other" rail side

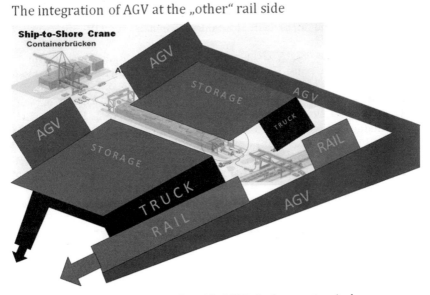

Fig. 6   The integration of rail transfer with AGV's in deep sea terminals.
Source: Adapted by Kortschak after: Engelhardt, T. and Müller-Elsner, H. (2003). Ein Gigant mit Gehirn, in: GEO, 11, pp. 114–140, pp. 132–133.

rail terminal area on the outer side of the deep sea terminal. This is one of possible solutions which could still be deployed at new Weser Jade Deep Sea Terminal in Wilhelmshaven in Germany.

4. Conclusion

Up to now, the terms of trade between rails and roads in hinterland traffic are in favor of roads because rail access is not integrated in deep sea terminals. By extending AGVs to rail mounted gantry cranes and performing the transshipment of containers on rail flat cars, the competitive disadvantage of rails could be eliminated. Furthermore, global warming and climate change require competitive transport chains including rails for hinterland traffic. Moreover, China plans to build huge railway infrastructures with links to Singapore. A competitive railway link should be integrated in the terminal operations. The proposed deployment of AGVs promises an efficient solution.

References

1. Duinkerken, M.B., Deekker, R., Kurstjens, S.T.G.L., Ottjes, J.A. and Dellaert, N.P. (2006). Comparing transportation systems for inter-terminal transport at the Maasvlakte container terminals, in: OR Spectrum 28, pp. 469–493.
2. Franke, K.-P. (2008). A technical approach tot he Agile Port System, in: Koenings, R., Priemus, H. and Nijkamp, P. (eds.): The Future of Intermodal Freight Transport Operations, *Design and Policy*, 2008, pp. 135–151.
3. Howaldtswerke — Deutsche Werft, A.G. *et al.* (1997). Container-Transportsysteme der Zukunft, Projekt, gefördert vom Bundesminister für Bildung, Wissenschaft, Forschung und Technologie, Förderkennzeichen 18S0071 A, Kiel.
4. Kortschak, B.H. (1992). CARGO NET-Lean Production for Combined Transport, in: CARGO SYSTEMS (ed.) Conference Proceedings, Intermodal 92, The Hague, pp. 77–78.
5. Muller, G. (1995). Intermodal Freight Transportation, 3rd ed., Landsdowne VA.
6. Vickermann, M.J. (1999). Agile Port Concept, http://www.transystems.com.

CHAPTER 10

ON THE ONGOING INCREASE OF CONTAINERSHIP SIZE

Simme Veldman

Ecorys Nederland BV

P.O. Box 4175, 3006 AD Rotterdam, Netherlands

simme.veldman@ecorys.com

For port planning purposes the development of the size of containerships is of great importance. Parties involved continuously try to beat competitors by creating the possibility to accommodate ships bigger than existing ones. With information available on vessel particulars and new building prices for fully cellular Post-Panamax container ships reaching up to about 14,000 TEU it is possible to conduct a proper statistical analysis of economies of ship size. Results show that economies of ship size, expressed as the elasticity of costs as a function of ship size, differ only slightly from those of ships up to Panamax.

The assessment of economies of ship size is similar to the one used by Ryder and Chappel (1979), Jansson and Shneerson (1987) and Cullinane and Khanna (1999) and (2000) and results show similarities and differences. Wijnolst *et al.* (1999) created a push in thinking about the role of much bigger ships with their Suez-Max and Malacca-Max designs and results are compared.

Economies of ships size act as the motor of demand for bigger ships. In order not to lead to too high user costs the increase in size has to be in balance with the combined increase in trade volumes and the number of port pairs between coast lines to be connected. The conclusion is that the ongoing increase in ship size will continue.

1. Introduction

The development of the size of containers ships is of great importance for port planning and the authorities involved try to beat competitors by creating the possibility to accommodate bigger, more cost effective ships. Since their introduction the size of containerships was increasing steadily up to Panamax and beyond. In this development limiting factors with the design of ships such as structural strength, engine capacity, cavitation of propeller and rudder, cargo handling equipment speed and available depth in ports were gradually solved and barriers pushed forward; this also applies to limiting factors in ports and their hinterland with respect to stock carrying and connecting inland transport.

Costs of capital and energy increase with the size of ship, but less than proportional and the relation between these costs and ship size can be well described by multiplicative relationships. The increase in costs as a function of ship size doubling from 500 TEU to 1000 TEU, then to 2000 TEU and so on to Panamax size of about 5000 TEU can be well described by these relationships and in log-linear form tested with regression analysis. This results in elasticities of ship size per category of costs, measured over a wide range of sizes of ships, showing remarkable similarities at both the lower and higher end of the range of ship sizes. See section on economies of ship size.

Do these economies of ship size also apply to Post Panamax ships, and to what degree? Results of a statistical analysis show the existence of economies of ships size for containerships ranging from 800 TEU to 14,000 TEU, of which new-building costs and engine capacity are known and also for the subsets of ships up to Panamax and Post Panamax separately. See Section on shipping costs of Post-Panama ships.

Will economies of ship size continue beyond 14,000 TEU? It is argued here that there are no a priori reasons to believe that the gradual process will stop with a size of, say, 20,000 TEU. Economies of ship size will keep the engineers and port planners busy in a stepwise manner, sometimes with small steps and sometimes with large bold ones, to solve the problems posed by the limiting factors as they did so far and thereby continue to do so in the future.

Containerships experience economies of ship size in their hauling capacity and diseconomies of scale in their handling capacity. To maximise scale effects it is therefore logic to employ big ships on the longest distance routes. The combined effect is that that the largest ships can best be employed on the North Europe — Far East trade route and our analysis therefore concentrates on the outcome on that route.

With a given volume of trade the employment of bigger ships implies that less roundtrips are needed to meet demand, leading to a lower frequency of service and thereby to higher stock carrying costs for shippers and receivers. This effect has a downward impact on the introduction of bigger ships.

An analysis of the introduction of the biggest ships shows that it takes place at a rate somewhat less than half the rate of the increase of trade, leaving the other half to an increase in the number of roundtrips. The introduction is negatively influenced by user costs and by the increase in ports to be covered per coast line, both requiring more trips to be made. The introduction is positively influenced by transhipment. See section on the ongoing increase in container ship size.

2. Economies of Ship Size

2.1. *Modeling ship size economies*

Ship size economies can be expressed as the shipping costs per unit of service as a function of ship size. Jansson and Shneerson (1987) developed an analytical framework based on multiplicative relationships between shipping cost categories such as capital, labour, energy and ship size and performance. They brought attention to the existence of economies of ship size with respect to a containership's hauling capacity and the diseconomies of its handling capacity. The ship size economies are based on specific technical relationships existing for capital related costs (construction, maintenance and insurance), manning costs and fuel use and fuel costs and related to the time spend at sea, while the handling capacity relates to the time spend in port. The statistical analysis hereafter is based on technical information and historic vessel prices of the existing fleet and order book of fully cellular containerships from WSE (2008) as it existed per January 2008.

2.2. *Capital related costs*

Capital related costs are based on the ship's new-building costs, which are the basis for ship market prices. Annual capital costs are set equal to the capital recovery factor (annuity) based on the interest rate and the ship's economic lifetime, where the interest rate is the weighed average of the return on equity and loans. The ship's price can be expressed as function of the ship's size and speed or size only according to:

$$P = \alpha_0 S^{\alpha 1} \tag{1a}$$

$$P = \alpha_0 S^{\alpha 1} V^{\alpha 2} \tag{1b}$$

where the price P is a function of size S, expressed in dwt or in TEU, and speed V in knots. The Greek letter symbols concern the coefficients of a multiplicative function and are estimated with regression analysis of the model in log-linear form. A value of α_1, to be referred to as elasticity, of less than one indicates that the ship's price increases less than proportionally with its size, implying that there are economies of ship size with respect to the ship's price and derived from that to capital and capital related costs.

For the regression analysis we use WSE (2008) containership data containing a great number of price quotations over a longer period of time. By using vessel prices over different years we have to allow for changes in time and do so by including yearly dummy variables, where $D_1, D_2 \ldots D_t$ refer to a dummy variable for year $1, 2 \ldots t$ compared to an arbitrary base year 0, for which 2008 is taken:

$$P = \alpha_0 S^{\alpha_1} V^{\alpha_2} \exp(\alpha_3 D_1 + \alpha_4 D_2 \ldots \alpha_t D_t) \qquad (1c)$$

In log-linear form this becomes:

$$\mathrm{Ln(P)} = \ln(\alpha_0) + \alpha_1 \ln(S) + \alpha_2 \ln(V) + \alpha_3 D_1 + \alpha_4 D_2 \ldots \alpha_t D_t \qquad (1d)$$

A higher design service speed, ceteris paribus, requires a greater engine capacity resulting in higher building costs. Bigger container ships appear to have a higher speed and the following multiplicative relationship appears to do well:

$$V = \beta_0 S^{\beta_1} \qquad (2)$$

A higher speed implies that there is a correlation between the two explanatory variables of model (1b), so that the regression analysis might suffer from multi-collinearity implying that there is a risk of obtaining erratic values of the estimates.

Ryder and Chappell (1979) did some measurements for containerships using quotations for a sample of the then existing fleet including containerships ranging from 80 to 580 TEU and published values of 0.48 and 1.38 for the size and speed coefficients respectively according to model (1b). The value of 0.48 is considerably smaller than one, suggesting the existence of great scale effects. For model (1a) they present a coefficient of 0.62. They did not publish goodness of fit indicators such as standard errors or R-square. The implicit relationship between the speed of ships and ship size according to model (2) can be assessed by equalling model (1a) and (1b) and solving for V, which leads to an elasticity of speed with respect to size according to β_1 of 0.19.

Jansson and Shneerson (1987) did a number of regression analyses for tankers, bulk carriers and containerships. They conclude that the ship capital cost is proportional to the two-thirds power of the ship size and apply this value for their analysis of containerships ranging from 240 to 3200 TEU. They did a number of regressions for general cargo ships and measured values ranging from 0.16 to 0.17 for different samples, which comes close to the value of β_2 of 0.19 as mentioned above.

Cullinane and Khanna (1999) tested model (1a) with 153 observations of a Fairplay database of container ships ranging from 1000 to 8000 TEU and estimated a size elasticity of 0.759 with an r-square of 0.93. Ship size was measured in nominal TEU.

We tested model (1b) for a sample of 1364 ships (to be) delivered in the period 1991–2012 of fully cellular containerships in excess of 10,000 dwt, which corresponds with about 800 TEU. The ship's size is measured in deadweight in order not to be influenced by playing by the owner with size in TEU. The spread in the size of ships is large enough to do statistical tests for fully cellular containerships up-to-Panamax and Post Panamax separately.

The statistical results of the first three lines in Table 1 show that estimated values of ship size elasticities according to model (1a) are statistically significant with high t-values for all data sets. For the set of all ships the value is 0.726 and for up to Panamax and Post-Panamax 0.677 and 0.745 respectively. The r-square values are high and t-values also, except for the speed variable. For the whole set the speed elasticity is 0.235 with a t-value of 2.6, meaning that the relation between design service speed and the ship's price is positive and significant. This also applies for the subset of ships up to Panamax with a coefficient value of 0.249 and a t-value of 2.3. For Post Panamax ships, however, the coefficient value is not significant

Table 1   Statistical test results of container ship price as a function of ship size and speed.

Dependent variable	Ln(dwt)		Ln(speed)		Sample description	DF	r-square
	α_1	t-value	α_2	t-value			
Ln(Price)	0.726	42.0	0.235	2.6	All ships	1364	0.895
Ln(Price)	0.677	29.8	0.249	2.3	Up to Panamax	1014	0.811
Ln(Price)	0.745	26.3	−0.290	−1.4	Post Panamax	349	0.910
Ln(Price)	0.766	92.4	Not included		All ships	1364	0.895
Ln(Price)	0.721	59.1	Not included		Up to Panamax	1014	0.810
Ln(Price)	0.733	27.2	Not included		Post Panamax	349	0.909

Estimated values of coefficients of yearly dummy variables have been omitted
DF: degrees of freedom
Source: Regression analysis based on WSE (2008) data.

and even shows a negative value. This erratic behaviour may be explained by the small variation in speed of Post Panamax ships.

If, according to model (1a) we only adopt ship size, the corresponding size elasticities would be 0.766 for all ships, 0.733 for Post Panamax and 0.721 for ships up to Panamax. The values of the ship size elasticities appear to be rather close and somewhat smaller for up to Panamax compared to Post Panamax. All values are close to the size elasticity of 0.759 estimated by Cullinane and Khanna (1999).

As stated there is a positive relationship between ship size and design service speed according to model (2). For the total set of containerships the elasticity is 0.167. For the subset of ships up to Panamax the elasticity is 0.174. For the subset of Post Panamax ships the elasticity is 0.029. It can be stated that the speed of ships increases with the size of ships until about Panamax and starts to be constant from there on. A scatter plot of size in TEU versus speed in knots is given in Fig. 1 and shows that the average design service speed of containerships ranges from about 16 knots for 800 TEU ships to 25 knots for up to Panamax ships. From there on it fluctuates at 25 knots. The fact that the design vessel speed does not

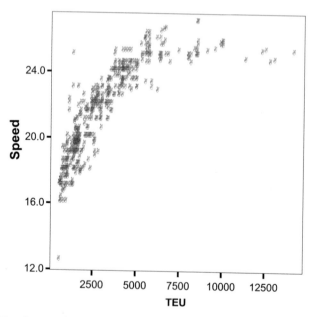

Fig. 1   Scatter diagram design service in knots as a function ship size in TEU based on WSE (2008) data.

Table 2   Statistical test results of design service speed as a function of ship size.

Dependent variable	Ln(dwt) β_1	t-value	Sample description	DF	r-square
Ln(Speed)	0.167	67.2	All ships	1364	0.768
Ln(Speed)	0.174	45.4	Up to Panamax	1014	0.670
Ln(Speed)	0.029	4.9	Post Panamax	349	0.061

Source: Regression analysis based on WSE (2008) data DF: degrees of freedom.

increase any more for Post Panamax ships is reflected in the poor statistical result for the speed variable for these ships.

For the calculations hereafter on economies of size for Post Panamax ships we take model (1a) as a basis using the coefficients as estimated for Post Panamax ships. A ship of 10,000 TEU, which corresponds with a tonnage of 123,000 dwt, has a price of USD 137 million in prices of 2008.

Jansson and Shneerson (1987) estimated for three different sets of general cargo ships with container capacity elasticity values ranging from 0.16 to 0.17. Cullinane and Khanna (1999) estimated an elasticity of 0.192 for a sample of 280 ships with an r-square of 0.90, somewhat higher than the 0.174 we found for ships up to Panamax.

Cullinane and Khanna (1999) assumed that operational costs such as costs of repairs and maintenance and ship insurance can be considered to be a fixed percentage of the ship's price thereby showing the same economies of ship size. Hereafter we will apply the same approach.

2.3. Labour related costs

Labour related costs depend on the size of the ship's crew, the nationality of the crew and the ship's voyage patterns. Small ships deployed in coastal shipping may have smaller crews and thereby lower costs of labour. These aspects have little to do with a ship's size. Including small ships Jansson and Shneerson (1987) measured a coefficient value according to model (3) of 0.03, close to zero; Cullinane and Khanna (1999) took two different amounts for crew costs for ships up to and in excess of 800 TEU respectively.

$$L(S) = \varepsilon_0 S^{\varepsilon 1} \tag{3}$$

In our analysis we concentrate on Post Panamax ships and therefore can take a fixed amount of USD 400,000 per year based on Drewry (2006).

Fuel related Costs

Fuel costs relate to a ships engine capacity. To attain a certain speed the horsepower requirement of the engine is less than proportional to ship size, which advantage is often traded off against higher speed. This effect can be expressed by a shipbuilding rule of thumb as mentioned by Jansson and Shneerson (1987) that horsepower (HP) is proportional to the two-third power of the displacement multiplied by the cube of the design speed. The equation can be expressed as:

$$HP = \gamma_0 S^{\gamma_1} V^{\gamma_2} \tag{4}$$

where capacity measured in horsepower (HP) is expressed as a function of ship size S and speed V. Of the set of WSE (2008) data used for the regression analysis with respect to a ship's price and speed, in 2% to 4% of the cases information on horsepower is missing. This small percentage most probably has little effect on the comparability of the outcome of the regression analyses on price and engine capacity.

The regression analysis results in Table 3 show that, depending on the set of ships, the estimated values range from 0.42 to 0.61 for the elasticity of engine capacity with respect to ship size and from 2.0 to 3.0 for the elasticity of engine capacity with respect to speed. The r-square value is the lowest for Post Panamax ships, which as noted before, most likely has to do with the small variation in speed.

The estimated values of the ship size and speed elasticities are smaller than those according to above-mentioned shipbuilding rule of thumb with values of 2/3 and 3. In their study Ryder and Chappell (1979) use values of 0.48 and 3.13 respectively. Jansson and Shneerson (1978) used a sample of ships of Zim Navigation Company and estimated an elasticity value of 0.72 for fuel consumption against ship size and did not include speed, as variations in design speed are minute.

Table 3   Statistical test results of main engine capacity as a function of ship size and speed.

Dependent variable	Ln(dwt)		Ln(speed)		Sample description	DF	r-square
	γ_1	t-value	γ_2	t-value			
Ln(HP)	0.607	42.6	2.215	29.7	All ships	1330	0.941
Ln(HP)	0.586	33.4	2.008	24.3	Up to Panamax	994	0.902
Ln(HP)	0.417	16.3	2.963	12.8	Post Panamax	335	0.624

Source: Regression analysis based on WSE (2008) data. DF: degrees of freedom.

Cullinane and Khanna (1999) estimated a value of 0.967 for ship size only. In their sample the speed of ships increases with size according to an elasticity of 0.192. Correcting for the increase in ship size the elasticity with respect to ship size comes at 0.78, which is still high compared to the values in Table 3.

The average consumption of heavy fuel oil (HFO) is 127 gram per brake horsepower for the 311 ships for with information on both horsepower and HFO consumption is available.

For the assessment of economies of ship size for Post Panamax ships we take model (4) as basis by using the estimated values of the coefficients. A ship of 10,000 TEU, which corresponds with a tonnage of 123,000 dwt and a design service speed of 25 knots, requires a main engine capacity of 85,700 HP. We assume that the design service speed can be achieved with 80% utilisation of the HP capacity. This means that the daily fuel consumption in tonnes comes at $85,700 \times 80\% \times 24/1000000 = 209$ tons. Cullinane and Khanna (1999) cite a publication of Gilman from 1980 stating that the daily costs of lubricating oil consumption are about 3% of HFO consumption. With a fuel price of USD 230 per tonne and 3% of costs of lubrication oils the daily costs when steaming come at USD 49,500.

Conclusions on Economies of Ship Size for Daily Costs

Comparing the estimated values of daily shipping cost elasticities of Post-Panamax size ships with ships up to Panamax it can be concluded:

- That for capital related costs the ships size elasticity is slightly higher or practically equal, meaning that economies of ships size are practically the same or slightly less.
- That for manning costs the ships size elasticity is also zero, meaning the economies of ship size are the same.
- That for fuel related costs the ship size elasticity is smaller, meaning that economies of ships size are greater.

3. Shipping Costs of Post Panamax Containerships

The basic elements of ship size economies relate to building costs, the costs of energy use and the costs of crews. Other costs such as those related to the handling of cargo and ships in port and the passage of canals are considered to be practically neutral with respect to the size of ships. Given the economies of ship size as estimated with regression analysis shipping

costs are assessed as a function of ship size for ships ranging from 6,000 to
14,000 TEU and extrapolated for ships reaching up to 20,000 TEU.

3.1. *Fixed annual costs*

The annual capital costs are assessed with a capital recovery factor (CRF)
of 10,19%, which is based on an interest rate of 8% and an economic lifetime
of 20 years. In line with Cullinane and Khanna (1999) costs of maintenance
and repairs, ship insurance and administration are set at 3.5% of the ship's
price, as it can be argued that these costs experience similar scale effects as
new-building costs. The sum of capital related costs appears to range from
USD 12.9 million for a 6,000 TEU ship to USD 31.3 million for a 20,000
TEU ship. The relation for Post-Panamax ships is based on model (1c),
while the design service speed of ships is expected to be 25 knots for the
whole range.

For Post Panamax ships crew costs do not vary with size and amount
to about USD 400,000 per year. The annual fixed costs by category and
size class are given in Table 4. The daily costs are based on a number of
350 operational days per year.

3.2. *Fuel costs*

The statistical test of model (4) shows that for Post Panamax ships the
main engine capacity increases with ship size with an elasticity of 0.417.
The design vessel speed of the Post Panamax ships does not increase with
size and is about 25 knots. The engine capacity varies with ship size from

Table 4   Fixed annual costs as a function of ships size in TEU (in USD 1000).

Size in TEU	Capital Costs	Maintenance, insurance and admin- istration	Total capital related costs	Manning and overhead	Total fixed costs	Total fixed costs in USD/day
6,000	9,627	3,308	12,935	400	13,335	38,099
8,000	11,887	4,085	15,971	400	16,371	46,775
10,000	13,999	4,810	18,809	400	19,209	54,884
12,000	16,000	5,498	21,499	400	21,899	62,568
14,000	17,914	6,156	24,070	400	24,470	69,916
16,000	19,757	6,789	26,546	400	26,946	76,987
18,000	21,538	7,401	28,939	400	29,339	83,826
20,000	23,267	7,995	31,263	400	31,663	90,465

Source: Based on outcome regression analysis as described in text.

Table 5   Daily cost at sea and in port (USD 1000).

Size in TEU	Daily fixed costs	Main engine capacity in HP	Daily fuel consumption tons	Daily fuel costs main engine	Daily fuel costs auxiliaries	Daily cost in port	Daily cost at sea
6,000	38.1	69,267	169	40.0	1.20	40,4	78.1
8,000	46.8	78,096	190	45.1	1.35	49,5	91.9
10,000	54.9	85,712	209	49.5	1.49	58,0	104.4
12,000	62.6	92,482	226	53.4	1.60	66,0	116.0
14,000	69.9	98,622	240	57.0	1.71	73,7	126.9
16,000	77.0	104,270	254	60.2	1.81	81,1	137.2
18,000	83.8	109,519	267	63.3	1.90	88,2	147.1
20,000	90.5	114,438	279	66.1	1.98	95,2	156.6

Source: Based on outcome regression analysis as described in text.

69,300 HP for a 6,000 dwt ship to 114,400 HP for a 20,000 TEU ship. The corresponding fuel consumption with 80% utilisation of the capacity ranges from 40 tonnes per day for the 6,000 TEU ship to 66.1 tonnes for a 20,000 TEU ship. With a fuel price of USD 230 per tonne the daily costs, including a 3% allowance for lubricating oil, range from USD 40,400 for the 6,000 TEU ship to USD 95,200 for the 20,000 TEU ship.

3.3. *Shipping costs per roundtrip*

Containerships experience economies of ship size in their hauling capacity and diseconomies of scale in their handling capacity. The hauling capacity is determined by size and sailing speed and the handling capacity by size and handling speed. The diseconomies of the latter are caused by the fact that, the larger the size of containerships, the longer the time to be spend in port. The time spent in port depends on the handling speed, which can be expressed as:

$$H = \varepsilon_0 S^{\varepsilon 1} \tag{5}$$

where handling speed H concerns the number of containers loaded and unloaded per day. The handling speed depends on factors such as crane productivity, the number of cranes working the ship simultaneously and the distribution of containers over the holds. As the speed is assessed over all ports called at on a roundtrip, some averaging takes place.

Jansson and Shneerson (1987) argued that the handling speed depends on the number of cranes used which should be proportional with the length of the ship. Results of regression analysis on WSE (2008) data show that this would lead to a value of 0.328 for the elasticity of cargo handling speed to ship size. Cullinane and Khanna (1999) elaborated on the subject, did a survey under operators resulting in a list of the number of cranes for varying sizes of ships and apply this for their calculations. Their list corresponds with an elasticity value of 0.47,[a] which implies that the cargo handling increases a bit more than with ship size than as according to Jansson and Shneerson (1987). For our calculations we take a conservative approach and apply the elasticity of 0.328.

Ships also need time to prepare for loading and unloading and for arriving and departure resulting in a fixed component of time per call.

For the assessment of the time ships spend in port the following assumptions are made:

- Roundtrip distance is $2 \times 12,000$ nautical miles with an average speed of 25 knots resulting in 40 days spent at sea.
- Per roundtrip 8 ports are called at with a fixed time per port of 6 hours resulting in 2 days.
- The number of laden and empty containers aboard is 90% of the carrying capacity resulting in $2 \times 90\% \times 20,000 = 36,000$ TEU for the biggest ship.
- The number of containers loaded and unloaded per roundtrip is based on a TEU/box ratio of 1.6 resulting in $36,000 \times 2/1.6 = 45,000$ boxes.
- The time needed for loading and unloading of containers for all ports depends on the product of the number of cranes working simultaneously and crane productivity. The average number of cranes working is put at 4 cranes for a 6,000 TEU ship and is assumed to increase proportionally with the ship's length leading to 5.9 cranes for the 20,000 TEU ship. Average crane productivity over all ports is put at 30 boxes per hour. For a 20,000 TEU containership requiring to have 45,000 moves per roundtrip, the time spend loading and unloading is $45,000/(5.9 \times 30 \times 24) = 10.6$ days.

Per ship size class the roundtrip time is given in Table 6 and increases from 46.7 days for the 6,000 TEU ship to 52.5 days for the 20,000 TEU ship. The total shipping costs of a roundtrip increase from USD 3.4 million to 7.6 million and per TEU carried the costs decrease from USD 314 to USD 207.

---

[a]The values presented in Fig. 4 of their article were measured from the graph and the elasticity value was measured with regression analysis.

Table 6 Cost and operational data per roundtrip.

Size in TEU 1,000	Time at sea in days	LOA in metes	No. cranes working	No. of containers handled 1,000	Handling time in days	Fixed time in days	Total port time in days	Roundtrip time in days	Total costs in USD 1,000	Containers shipped in TEU 1,000	Cost per TEU
6	40	291	4.0	13.5	4.7	2	6.7	46.7	3,448	10.8	314
8	40	320	4.4	18	5.7	2	7.7	47.7	4,131	14.4	282
10	40	344	4.7	22.5	6.6	2	8.6	48.6	4,771	18	260
12	40	365	5.0	27	7.5	2	9.5	49.5	5,381	21.6	244
14	40	384	5.3	31.5	8.3	2	10.3	50.3	5,969	25.2	231
16	40	401	5.5	36	9.1	2	11.1	51.1	6,540	28.8	222
18	40	417	5.7	40.5	9.8	2	11.8	51.8	7,099	32.4	214
20	40	432	5.9	45	10.5	2	12.5	52.5	7,647	36	207

Source: Based on outcome regression analysis as described in text.

Conclusions on Economies of Ship Size on Long Distance Trade

The statistical analysis shows that ship size elasticities of Post Panamax ships with respect to daily costs are 0.733 for capital related costs, 0.0 for labour and 0.417 for fuel. These elasticities have to be combined with the operational performance expressed in the elasticity of time spend at sea (0.0) and the elasticity of time spend in port (0.33). The combined effect results in total roundtrip costs as a function of ship size.

In order to compare the resulting economies of ship size with the outcome of other studies, we use the elasticities of shipping costs as a function of ship size defined as $(\delta c/c)/(\delta s/s)$, where c represents costs and s size. The total cost function is not multiplicative, so that the elasticity value varies with s and, in fact, decreases as the size of ships increases, which reflects the increasing importance of time spend in port. The elasticity values therefore apply to certain ship size intervals, for which it is assumed that the function is a multiplicative relationship according to:

$$TC = \zeta_0 S^{\zeta_1} \tag{6a}$$

$$TC/S = \zeta_0 S^{\zeta_1}/S = \zeta_0 S^{\zeta_1-1} \tag{6b}$$

where TC concerns the total costs of a roundtrip and TC/S the cost per TEU carrying capacity and, given a fixed load degree for all ships of the range, per TEU carried.

The value of the costs elasticity is 0.65 for the whole range of ships and is close to the economies of ships size with respect to capital costs, but higher than those for energy and crew costs. The number of containers carried is proportional to the size of ships, so that the costs per TEU carried have on average an elasticity with respect to ship size of $\zeta_1 - 1 = 0.65 - 1 = -0.35$. This means that a 1% increase in ship size will lead to a -0.35% decrease in shipping costs per TEU. Over the whole range the value of the elasticity varies from -0.38 for ships of 6,000 TEU at the lower end to -0.30 for ships of 20,000 TEU at the upper end of the range. Note that if the cargo handling efficiency increases in the course of time, the diseconomies related time spend in port becomes less important, which leads to greater economies of ship size.

From Fig. 2 can be seen that the shipping costs of the time spend at sea (hauling costs) decrease as a function of ship size and the cost of the time spend in port (handling costs) increase as a function of ship size. The diseconomies of ship size of the handling capacity practically appear not

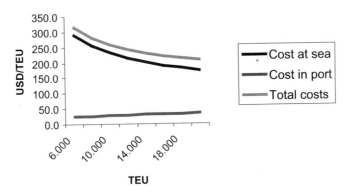

Fig. 2   Cost at sea, in port and total costs as a function of ship size.

to play a role for containerships ranging from 6,000 to 20,000 TEU with a roundtrip distance of 2 × 12,000 nautical miles. For the range of ship sizes studied the share of port time increases from 14% to 24% of the total roundtrip time, so that the diseconomies of the hauling capacity have little negative impact on economies of scale.

Jansson and Shneerson (1987) calculated shipping costs as a function of ships size for container ships ranging from 240 to 3200 TEU based on their assumptions on ship size economies as discussed above. At the time they wrote their book port productivity was substantially lower than now, so that on a roundtrip of 2 × 10,400 nautical miles the time spent in port was rather long. For the range of ships they studied the economies of scale as experienced with capital, fuel and labour costs after a certain size of ships were more than compensated by diseconomies experienced with the time spend in port. It appears that economies of ship size exist for the range of ships from 240 to 1800 TEU and correspond with an elasticity of ship size of −0.34 for ships at the lower end. From 1800 TEU on costs are practically constant turning in a slight increase per TEU.

Economies of scale as derived from the shipping costs as calculated by Cullinane and Khanna (1999) for a long distance route of 11,000 nautical miles for the range of ships from 1000 to 8000 TEU vary from −0.42 to −0.20 with an average of −0.25 for the whole range.

Wijnolst et al. (1999) studied comparable shipping costs for the route Singapore — Rotterdam for existing container ships and two new designs named Suez-Max and Malacca-Max. The related elasticities of ship size[b]

---

[b]Based on a multiplicative relationship between two consecutive cost observations.

appear to be −0.25 for the range from 10,000 to 15,000 TEU and −0.35 for the range from 15,000 to 18,000 TEU.

4. The Ongoing Increase of Containership Size

4.1. Development of ship size and trade

A certain degree of urgency is needed to stimulate technological improvements intended to realise economies of ship size. Expected high increases in demand of containerships stimulate ship owners to order, ship engine and crane builders to innovate and port authorities to accommodate for bigger ships. All this together makes it possible to realise economies of ship size.

Since their introduction in the late sixties the size of containerships is increasing gradually. Figure 3 shows that the size of the biggest ship in service increased from 4,600 TEU in 1988 to 12,500 TEU in 2006. If the ships on order as in January 2008 are included, the size of the biggest ships will increase further to 14,200 from 2011 onward. The increase in size of the biggest ship in operation shows some steep increases by trendsetting ordering by operators such as Maersk. The figure shows that the increase in size has a more gradual pattern, if the criterion for "biggest ship" is measured as the average size of the 2% of the biggest ships of the total fleet of fully cellular containerships.[c]

From 1988 to 2007 the size of ships, measured as the average of the biggest 2%, increased annually with 3.8%. The increase was 3% during first half and 4.8% during the second half of the period. An explanation of the lower rate in the first half most likely has to do with the impact of the

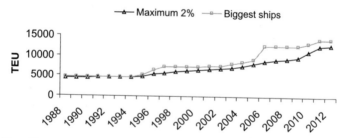

Fig. 3   Size of biggest ship and average size of 2% biggest ships per year.
Source: Derived from WSE (2008) Data.

---

[c]Note that the due to cancellation of orders the actual average size from 2008 on will be lower.

Panama Canal on ship size, when ordering Post Panamax ships gradually got momentum. Accelerations in the increase in ship size are from 1996 to 1998, from 2004 to 2006 and, if the order book as at January 2008 is taken, also from 2010 to 2012.

   The North Europe — Far East trade route has the longest distance between the world's major industrial areas and thereby is the one where the advantage of the biggest ships is best utilised. The 2% biggest ships in 2007 concern 87 units, which correspond with about 11 strings (defined a liner service providing one roundtrip per week) on the North Europe — Far East trade route assuming that 8 ships are needed to produce a string. For the assessment of the average TEU of the top 2% ships we took the fully cellular container fleet as per 1/1/2008 and on order as basis. This means that numbers of shipped scrapped between the year concerned and 2008 are disregarded. This means that the real number of ships of the fleet in that year is greater and thereby also the stated number of ships in the top 2% range. Less ships means that the calculated average will be somewhat higher than as in reality. It may be expected that this effect is little given the fast increase in the fleet of ships.

   Figures compiled from Drewry (1992), Drewry (2004) and Drewry (2007) show that the volume of containers carried on the North-Europe — Far East trade route increased from 2.9 million TEU in 1990 to 13.5 million TEU in 2007, which is in annual average increase of 9.6%. Over the same period the increase in size of ships was 4.3%.

   An analysis of information of shipping services as published by Drewry (1992) shows that in 1992 the equivalent of 15 weekly services were offered.

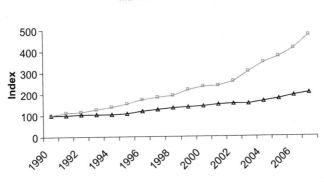

Fig. 4   Index volume North Europe — Far East trade route and average size 2% biggest containerships per year.

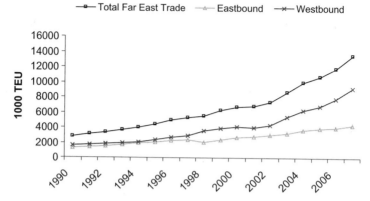

Fig. 5   North Europe — Far East Trade route volumes eastbound, westbound and total per year.
Source: compiled from Drewry (1992), Drewry (2004) and Drewry (2007).

Similar information by Drewry (2007) shows that there were 34 services offered in 2007. The corresponding annual growth rate is 5.6% annually. The combined annual growth factor from those roundtrips and ship size comes at $(1.056 \times 1.043) = 1.10$.

This annual increase of 10% per year corresponds rather well with the increase in total trade volume of the North Europe — Far East trade route, which is 9.6% for the total volume and 10.8% for the trade of the heaviest link, i.e., the westbound trade.

4.2. Factors limiting economies of ship size

The economies of ship size as measured in the preceding Section relate to the size of container ships up to about 14,000 TEU and are extrapolated. As in the past this extrapolation will meet limitations with respect to the possibility to continue with a single engine, with increasing cavitation problems of propeller and rudder, with trading draft in ports, with the dimensions of the New Panama Canal and, in general, with sufficient demand to fill the ships. These problems already existed in the past and have gradually been solved for ships up to 14,000 TEU.

4.2.1. Structural design, engine and cavitation problems

In a technical paper on the state of the art of big containerships Tozer (2006) states that there are no insurmountable problems with respect to structural design. With respect to problems associated with a single engine and cavitation problems with propeller and rudder Tozer (2006) mentions that propeller design will provide an upper bound on ship size or speed, rather than the availability of main engines to provide the necessary power. From the perspective of ship design it is at this stage not possible to assess a certain upper limit from where on a double set of engines and propellers are needed.

4.2.2. *Trading draft*

Draft in seaports is often put forward as a crucial barrier for economies of ship size. Tozer (2006) states that the trading draft of container ships generally is less than the maximum draft, as for the important trades the maximum weight as dictated by the market is up to about 90% of the maximum. An analysis of the relation between maximum draft and ship size for Post Panamax ships show that the increase in draft as a function of size is less than as for ships up to Panamax. The largest Post Panamax ships have a maximum draft of 16 meters for the biggest ships of Maersk. This effect is also reflected in the value of the elasticity of maximum draft with respect to ship size in TEU, which has a low value of 0.168 for Post Panamax ships against 0.272 for ships up to Panamax. The effect suggests that for the bigger ships a kind of shallow draft design applies intended to better deal with problems of draft.

The "shallow draft" design effect as measured for Post Panamax ships from 6,000 TEU to about 14,000 TEU can be extrapolated to the size class of the Malacca-Max ships of 18,000 TEU. Assuming that the shallow draft design will continue to exist to the same degree, we will continue to apply a draft versus size elasticity of 1/3. This would result in a maximum draft of about $16 \times (18{,}000/13{,}000)^{\wedge}(1/3) = 17.8$ meters instead of 21 meters as designed for the Malacca-Max containership.

It should also be noted that the maximum draft of 16 meters will not be reached often, given the difference between maximum draft and trading draft on the major east-west trade routes. An example as to how far the trading draft may differ from the actual draft in a systematic manner concerns the allocation of the MSC Beatrice of 13,800 TEU with a maximum draft of 14.5 meters on a service calling at Antwerp with a considerably lower maximum depth.[d]

---

[d]Containerisation International Monthly, May 2009, p. 29.

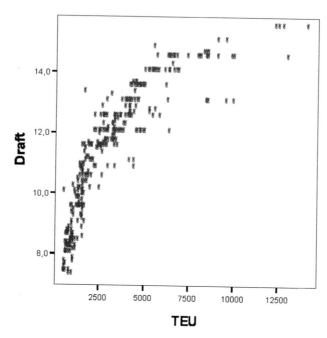

Fig. 6   Draft versus ship size.
Source: WSE 2008 data.

Table 7   Existing and new Panama Canal restrictions for ships passing.

Vessel dimensions (in meters)	Existing Panama Canal	New Panama Canal	% increase
Length over all (LOA)	295	426.7	45%
Beam	32.2	55.9	74%
Draft	12.0	18.3	53%

Source: Wikipedia (2009).

4.2.3. *The New Panama Canal*

In the past the design of containerships was strongly influenced by the limitations of the Panama Canal with dimensions as given in Table 7. Will this happen again and how will this influence the design of containerships? With a maximum of 32.2 meters the vessel's beam posed the greatest restriction rather than its LOA maximum of 295 meters. With a beam restriction of 55.9 meters the New Panama Canal (NPC) allows for the passage of ships with a 74% greater beam. At present the maximum size of a Panamax containership given the Panamax beam restriction of 32.2 meters is about 5,000 TEU.

With a beam of 56.6 meters the biggest ships of Maersk are just below the maximum beam of 55.9 meters. The biggest ships on order by companies other than Maersk range in size from 13,300 to 14,200 TEU and all have a beam of less than 50 meters and a LOA of 366 meters. These observations suggest that the maximum TEU of ships with dimensions optimised with respect to the NPC are considerably greater than 15,000 TEU, although it is not known how much bigger.

Of the main world trade routes the deployment of ships on the TransAtlantic and TransPacific is affected by the dimensions of the Panama Canal. These restrictions lead to use of land-bridges extending the TransAtlantic to the west of North America and the Transpacific to the east part. This will happen again with the New Panama Canal and what will be the size of the NewPanamax design? This will affect the increase in the size of containerships.

4.3. A balance between user and producer costs

In the previous Section the conclusion was made, that economies of ship size as they exist for capital, fuel and crewing costs for ship up to Panamax, continue for Post Panamax ships. The combined effect of economies of ship size of the hauling capacity and diseconomies of ship size of the handling capacity results in a elasticity of shipping cost per container shipped on the Europe — Far East trade of -0.35, i.e., a small increase in ship size of 1%, leads to a reduction of shipping costs of -0.35% and being a bit less for the ships of about 20,000 TEU. It was further concluded that technical limitations with respect to increases in ship size beyond the size of the presently largest ships of about 14,000 TEU most probably are still elastic, as they were in the past. It is, however, unsure how far that goes.

The optimum size of ships is not only determined by shipping costs as incurred by the ship operator, the producer of shipping services. The costs incurred by the users of these services, the shippers and receivers of cargo, also may play a role. Jansson and Shneerson (1987) state, that the optimum size of ships on a trade is determined by the sum of user and producer costs. Producer costs are the costs needed for the production of shipping services as discussed before, while user costs concern all other logistic costs incurred by the user of a shipping service such as stock carrying costs. The employment of bigger ships results in less roundtrips to be made, given a fixed volume of trade. This leads to a longer time span between two consecutive sailings resulting in more storage costs incurred by users. This aspect is one of the factors limiting the size of ships on many routes: employment of bigger ships leads to lower producer costs, but the users may not like the resulting lower frequency of service. These effects are referred to as economies of ship size of producer costs and diseconomies of ship size related to numbers, i.e., related to the number of roundtrips.

Veldman (1993) shows on the basis of cross-section data for 1987 that the average size of ships employed on the shipping routes connecting the world maritime regions can be explained by trade level and sailing distance.

The size of containerships employed on secondary container routes is considerably smaller than the maximum size of ships and the cargo volume offered on these route is not sufficient to load big ships with a sufficiently high frequency of service. In other words, the users of the shipping services on these routes demand a frequency of service which provides a sufficiently high load factor for the ships employed.

As mentioned above Jansson and Shneerson (1987) worked out an example of a liner shipping route demonstrating the trade-off between economies of ship size at sea and diseconomies of ship size in port for vessels ranging from 240–3200 TEU. With the low cargo handling efficiency of that time the optimum ship size was about 2400 TEU. By adding stock carrying and interest costs the minimum of the total of user and producer costs is reached for much smaller size classes. Veldman (1993) gave an example of the sum of user and producer costs for the Europe — Far East trade route, demonstrating how the total cost curve is U-shaped for the interval of ships ranging from 1000 to 4500 TEU and varies as a function of the value of trade.

The issue of stock carrying costs plays a role in cases where the number of roundtrips offered is below what the market is asking. What is the market? An operator investing in new larger tonnage in order to reduce its costs to become more competitive has to pay attention to the frequency of service he is offering for the port pairs connected by his service. This means that most likely he will not deploy, given his market share, much bigger ships so that he has to offer fewer services. Instead, he will instead offer at least the same number of services/roundtrips, with bigger ships. Then, the increase in ship size will reflect his opinion about the increase of the total market and his market share therein. This means that in a fast growing market he will take bigger steps than in a slow growing one. The big new-building bonanza of the last years is proof of this.

4.4. *Factors affecting development of user costs*

4.4.1. *Numbers of ports being connected*

The shipping services offered on a trade route are a mixture of hub-and-spoke operations and multi-porting operations. On the Europe — Far East trade route ships may call at about 5 ports at the European end and at 6 ports at the Asian end. As at the Asian end they may call at some ports twice, inbound and outbound, the actual number of ports called at in Asia may rather be 5. The itinerary patterns of course differ per operator, some of who rely more on transhipment than others.

OSC (2006a) and OSC (2006b) present historic port throughput figures, which can be aggregated by coastal areas. From these figures it appears that the number of ports for the coastal areas served by the Europe — Far East trade route, there is an increase in the number of ports. A broad analysis suggests that over the period 1990–2006 the increase was about 50% per coastal area. These ports include mainline, feeder line and regional ports and it is likely that most of them are linked to the worldwide container network. Roughly, the number of ports connected increase with a factor $1.5 \times 1.5 = 2.25$, which corresponds with an annual rate of 5 percent. This increase in the number of port-pairs to be connected puts the increase of the number of roundtrips of 5.6%, as mentioned in Section 4.1 in a different light. It suggests that the increase in roundtrips is needed to compensate for the increase in port-pairs to be served in order to maintain the frequency of service between ports.

Another way to deal with the increasing number of port pairs is transhipment. On a typical service the number of port pairs to be connected comes at $5 \times 5 = 25$ on a total of, say, $50 \times 50 = 2500$ different port pairs to be connected on the West Europe — Far East trade route. If an operator would like to cover all these port pairs he would need to offer $2500/25 = 100$ different services. This is much and it may be clear that some degree of transhipment is inevitable.

4.4.2. *Development of transhipment*

Hub-and-spoke (HS) operations are a means to increase the coverage of main lines serving two coastlines thereby also serving the port pairs connecting the coastlines that are not linked directly by multi-porting itineraries. To make HS operations competitive with multi-porting operations, given a certain market size, it is required that the mainline part of the costs is as low as possible by deployment of the largest possible ships. Cost comparisons with a model by Jansson and Sheerson (1987) show that, given a fixed market, a HS system is only under exceptional circumstances more cost effective than a multi-porting system. This concerns situations where the size of the mainline ship deployed in the HS system is big in comparison to the ship of the multi-porting system, where cargo handling costs are low and the geographical setting favourable.

Under the heading "Direct call versus transhipment debate" Baird (2005) mentions the general opinion under maritime economists of the view that extra feeder and handling costs make transhipment more expensive than multi-porting. This supports the findings of Jansson and Shneerson (1987).

Not all transhipment has to do with hub-and-spoke operations. Inter-liner transhipment, linking one mainline with the other, does not and generally takes place in certain specialised ports favourably located at cross points of east-west and north-south trade routes, such as Algeciras, Salalah and Singapore. Inter-liner transhipment as far as it takes place within a trade route however is a means to increase the coverage of port pairs on that trade route in a similar way as HS operations.

HS operations are an extra reason, i.e., additional to the effort to just beat each other with bigger ships, to stimulate the introduction of bigger ships. The introduction of bigger ships itself seems to stimulate HS operations as the number of ports able to accommodate such ships is less than for the ships they replace. Introduction of HS operations and big ships seems to mutually strengthen each other.

How is transhipment developing: is it increasing or decreasing? No information is available at the level of container flows, but there is at the level of port throughput. A compilation of data from Drewry (2004), (2007 and 2008) shows that the share of transhipment in total container throughput at global level is increasing gradually from 11% in 1980 to 18% in 1990, to 25% in 2000 and further to 27% in 2007. Looking at the development since 2001 it appears that growth has stopped.

Per coastal region the development is different. For North Europe the transhipment share is increasing from 14% in 1980 to 18% in 1990 further increasing to 24% in 2002 and fluctuating below 24% since then. The geography of South Europe is more favourable for transhipment, while containerisation started later: the share of transhipment increased from 15% in 1980 to 25% in 1990 and further to 35% in 2000 and is increasing still further to 43% in 2007.

Table 8   Development of the share of transhipment for coastal regions.

Region	1980	1990	2000	2001	2002	2003	2004	2005	2006	2007
N. Europe	14%	18%	20%	23%	24%	24%	21%	22%	23%	24%
S. Europe	15%	25%	35%	35%	37%	36%	37%	42%	43%	43%
SE Asia	32%	40%	48%	46%	46%	47%	48%	50%	50%	51%
Far East	12%	18%	23%	26%	27%	26%	23%	22%	21%	21%
Total world	11%	18%	25%	27%	27%	27%	26%	27%	27%	27%

Sources: Compiled from Drewry (2004), Drewry (2007) and Drewry (2008).
Note: In case figures in different publications differ, those of the newest one are used.

The region Far East shows an increase from 12% in 1980 to 18% in 1990 and further to 23% in 2000 reaching a peak of 27% in 2003. Since then it is decreasing gradually to 21% in 2007. The geography of Southeast Asia is favourable for transhipment. From a high level of 32% in 1980 it is gradually increasing to 40% in 1990, 48% in 2000 and further to 51% in 2007.

The long term development of the share of transhipment from 1980 to 2007 suggests that at a global level transhipment has reached its ceiling. Of the coastline regions dominating the Europe — Fareast trade route the Far East and West Europe seem to have reached a ceiling, while for the Far East it has turned into a slight decrease. South Europe and Southeast Asia still experience some growth.

It can be concluded that the increase in trade on the North Europe — Far East trade route results in an increase in ship size and an increase in number of roundtrips. The increase in the number of port pairs to be connected seems to be met by an increase in roundtrips in combination with an increase in HS operations. Since the first years of the century the increase of transhipment and thereby of HS operations, seems to have reached a ceiling.

5. Conclusions

The results of the statistical analysis based on a great sample of containerships shows that economies of ship size forz Post-Panamax ships, as expressed in elasticities of costs as a function of ship size, differ only slightly from those of ships up to Panamax size. Economies of ship size are reflected in the resulting shipping costs on the North Europe — Far East shipping route.

The values of the estimated elasticities are similar to those of earlier research concerning smaller ships and lower cargo handling efficiency in ports such as by Ryder and Chappel (1979) for ships ranging from 80 to 580 TEU, by Jansson and Shneerson (1987) for ships ranging from 240 to 3200 TEU and by Cullinane and Khanna (1999) for ships ranging from 1000 to 8000 TEU. The results suggest that scale effects are equal or slightly less for capital related costs and slightly higher for fuel costs.

Economies of ships size act as the motor of demand of bigger ships. In order not to lead to too high user costs the increase in size has to be in balance with the increase in trade volumes, in the number of port pairs to be served, the number of roundtrips and in the share of transhipment operations.

The technical limitations to the size ships are gradually being pushed further. The analysis shows that there are good arguments to extrapolate to the scale effects as measured for ships up to 14,000 TEU further to ship sizes up to 20,000 TEU. The observations suggest that the ongoing increase in ship size will continue.

The development shows that over the period 1990 to 2007 the increase in trade on the North Europe — Far East trade route in west bound direction, the motor of the increase in ship size, increased with 10.8% per year, while the increase in the size of the top 2% containerships increased with from 4,526 TEU to 9,220 TEU, corresponding with an annual growth rate of 4.3%. Over the same period the number of weekly services offered on the trade route increased with 5.6% annually, matching fairly well the difference of the increase in ship size and the increase in trade.

The number of ports located along the coastlines at the end of the trade route based on OSC (2006) is increasing and rough estimates suggest with a rate of about 5% annually. To cope with this increased density of port pairs on the trade route more roundtrips need to be made in combination with an increase in transhipment. The 5.6% increase in the number of roundtrips suggests that this is the major factor, while the increase in transhipment, which seems to have stopped short after 2000, appears to have become a less important factor.

With the motor of economies of scale continuing to exist for ships in excess of 14,000 TEU there are good reasons to expect that the average size of the top 2% ships will increase further at a rate somewhat lower than as for the increase in demand of trade as it did in the past. Part of this effect has already taken place as reflected in the number of big ships in the order book as per 1/1/2008.

The observed stop of the growth of transhipment may have a slight downward impact on the ratio of the growth rate of ships size versus demand, i.e., 4.3% versus 10.8%. This effect, however, may be compensated by a stop in the increase in seaports per coastal area, which has a downward impact.

So far the long run development. Since the end of 2008 the increase in worldwide container trade and in the North Europe — Far East trade has come to a standstill and turned into a decrease, as a result of the credit crunch in combination with overoptimistic ordering. It will take time for the world economy to recover and to reach the 2008 level again, for world trade to recover and reach the 2008 level again and when the increase of trade will have started again to reach levels where the existing over-tonnage will have been worked away. If that moment has arrived ordering will start again and also ordering of container ships in excess of the biggest ships existing now and now on order.

References

1. Baird, A. (2005). "Optimising the container transhipment hub location in northern Europe", *Journal of Transport Geography*, Article in Press, www.sciencedirect.com.
2. Cullinane, K. and Khanna, M. (1999). Economies of scale in large container ships, *Journal of Transport Economics and Policy*, 33, pp. 185–208.
3. Cullinane, K. and Khanna, M. (2000). Economies of scale in large containerships: optimal size and geographical implications, *Journal of Transport Geography*, 8, pp. 181–195.
4. Drewry Shipping Consultants (1992). Strategy and Profitability in Global Container Shipping.
5. Drewry Shipping Consultants (2004). Container Container Market Review and Forecast — 2004/05.
6. Drewry Shipping Consultants (2006). Ship Operating Costs, Annual Review and Forecast — 2006/07.
7. Drewry Shipping Consultants (2007). Container Container Market Review and Forecast — 2007/08.
8. Drewry Shipping Consultants (2008). Container Forecaster Annual.
9. Jansson, J.O. and Shneerson, D. (1987). *Liner Shipping Economics*, Section 5.5, Chapman and Hall Ltd, London.
10. Ocean Shipping Consultants (OSC) (2006a). The European & Mediterranean Containerport Markets to 2015.
11. Ocean Shipping Consultants (OSC) (2006b). East Asian Container Port Markets to 2020.
12. Ryder, S. and Chappell, D. (1979). *Optimal Speed and Ship Size for the Liner Trades*, Marime Transport Centre, Liverpool.

13. Tozer, D. (2006). Design challenges of large container ships, Lloyd's Register, Paper presented at ICHCA.

14. Veldman, S. (1993). The optimum size of ships and the impact of user costs, an application to container shipping, Chapter 9 in *Current Issues in Maritime Economics*, Edited by K.M. Gwilliam, Kluwer Academic Publishers.

15. WES (2008). World Shipping Encyclopaedia of Lloyd's/Fairplay.

16. Wijnolst, N., Scholtens, M. and Waals, F. (1999). *Malacca-Max, The Ultimate Container Carriers*, Delft University Press.

CHAPTER 11

A LINEARIZED APPROACH FOR LINER SHIP FLEET PLANNING WITH DEMAND UNCERTAINTY

Qiang Meng*, Tingsong Wang and Shahin Gelareh
*Department of Civil Engineering, National University of Singapore
P.O. Box 117576, Singapore
* cvemq@nus.edu.sg*

In this paper, we focus on the liner Ship fleet planning (LSFP) with cargo demand uncertainty. The LSFP aims to determine which type of ships and how many of them are needed, and how to deploy and operate these ships. We first propose a mixed integer nonlinear programming model for the LSFP by taking into account cargo shipment demand uncertainty. We find that the fuel consumption cost of a ship can be approximated by a linear function with respect to its cruising speed, and we then proceed to build a mixed integer linear programming model that can approximate the originally proposed nonlinear programming model. The mixed integer programming model can be thus effectively solved by some optimization solvers. Finally, numerical examples are carried out to assess the linearized approach.

1. Introduction

1.1. Background

Due to the increased global trade activities, the maritime transportation industry has been growing steadily during the past decade. In particular, highly-containerized liner trade is the fastest growing sector. Liner shipping occupies the most major place in the global trading transportation. UNCTAD[1] reported that the share of the top 20 liner operators held about 70 percent of the total container capacity deployed in 2008, and it pointed out that with the continuous advancement of ship building technology and incensement of global container traffic, the liner shipping dominant trend would continue to strengthen.

Since the ships are assets with expensive operational cost, how to effectively manage the ship fleet in order to maximize its profit or minimize its cost is an issue highly concerned by the liner operators. This issue is termed as a Liner Ship Fleet Planning (LSFP) problem in this paper. There are two tasks involved in LSFP problem: one is the planning of fleet size and mix; and the other is the deployment of the fleet. The fleet size refers to the total number of ships needed to ship cargoes for shippers, while the mix refers to the types of ships in the fleet. For example, Oriental Oversea Container Line (OOCL), a liner shipping company in Hong Kong, owns a fleet with the size of 39 ships constituted of 9 different types in 2009.[2] Fleet deployment is the ship-to-route allocation. We will deal with these two tasks simultaneously in this paper and aim at developing a mathematical programming model in order to assist a liner operator to manage their ship fleet effectively, thereby operating the fleet with a cost as less as possible. The core of this model is a stochastic programming model where the demand of containers delivered between any two ports is considered to be random variablea. It is firstly formulated as a mixed-integer nonlinear programming (MINLP) model with consideration of ships' cruising speed as decision variables. We find that the fuel consumption cost for a ship can be approximated by a linear function with respect to its cruising speed, and we then proceed to build a mixed integer linear programming (MILP) model to approximate nonlinear programming model. Finally, the solver CPLEX is employed to solve the MILP model.

1.2. Literature review

There are a number of studies focusing on ship fleet planning problem, most of which are surveyed in the following four major review articles or chapter: Ronen,[3,4] Perakis[5] and Christiansen et al.[6]

To the best knowledge of authors, this paper is first to incorporate uncertain demand into optimization model for LSFP. Thus, the literature directly related to this paper is focused on those dealing with deterministic demand, which falls into two categories: modeling for fleet's size and mix problem and for fleet deployment problem.

The decision-making of liner ship fleet size and mix is one of the important tasks in LSFP for both researchers and shipping service operators. Some related researches have been studied: Dantzig and Fulkerson[7] addressed the fleet's size planning problem for a tanker fleet, Laderman *et al.*[8] discussed the same problem for a tramp or an industrial shipping company and Everett *et al.*[9] proposed a linear programming model for large bulkers and tankers fleet. A simple approach was proposed by Benford[10] to select the optimal mix of available bulkers and cruising speed operated between two specific ports. Afterwards, Perakis[11] solved the problem of Benford[10] by using Lagrangian multipliers approach.

However, those researches studied are either for tramp ships or industrial ships, not for liner ships. Liner ships are different with the first two shipping modes. It provides a fixed liner service, at regular intervals, between named ports and offers transport to any shippers. In liner shipping, time is very important due to the fact that liner ships have to comply with the schedules even the operation is at low utilization levels (Lawrence;[12] Stopford.[13]) Lane *et al.*[14] developed a three phased approach to determine the fleet size and mix for a liner trade route with known demands for shipping services between any origin-and-destination pairs. A three-phase approach was developed by Fagerholt[15] to find the optimal fleet design. Recently, Hsu and Hsieh[16] formulated a two-objective model to determine the optimal liner routing, ship size and sailing frequency of container carriers by using the Pareto principle. Tabu search algorithm is applied by Brandao[17] to seek the optimal fleet size and mix planning for vehicle routing problem.

Some literatures studied the fleet deployment problem, which is another important task in LSFP: Perakis and Papadakis[18,19] proposed a mathematical model to determine the optimal fleet deployment and sensitivity analysis was performed to study the effects of small or large changes in one or more cost components on the total cost. Later, the multiorigin, multidestination fleet deployment problem was studied by Papadakis and Perakis.[20] In this paper, the speeds of vessels were categorized into two types: full load and ballast speeds. A projected Lagrangian method was applied to solve the problem. A realistic model for optimal deployment with detailed description of operating costs of liner ships was proposed by Perakis and Jaramillo,[21] and their proposed model is solved by LINDO solver in Jaramillo and Perakis.[22] Powell and Perakis[23] revisited the problem studied by Perakis and Jaramillo[21] and formulated it as an integer programming model. Millar and Gunn[24] formulated two mixed-integer programming models to dispatch a fishing trawler fleet and used a heuristic method to solve them. Vukadinović and Teodorović[25] discussed the dispatching problem of barges by using the fuzzy approach. A linear programming model and a mixed integer programming model by Cho and Perakis[26] investigated liner fleet deployment and expressed it as matrix forms by introducing flow-route incidence matrix. A mathematical program solver OSL was employed to solve their model. Mourão et al.[27] presented an application of integer programming model to support the decision-making process of assigning ships with hub and spoke constraints.

Those researches reviewed in the above sections are all for short-term planning horizon, for the long-term planning, Nicholson and Pullen[28] did the first study on fleet management of sale and replacement and used a two-stage method to tackle it. Xie et al.[29] applied a dynamic programming model to seek the best liner fleet deployment policy for long-term planning horizon, and for short-term planning horizon (each one year), it was formulated as a linear integer programming model. The problem was solved by a heuristic algorithm.

1.3. Randomness

There is one major source of randomness in the LSFP problem, namely the cargo shipment demand. Before describing it, we first briefly introduce the procedures of a typical shipping flow as follows: a shipper books space from a liner shipping company through a shipping agent at port of shipment by filling in a shipping application (S/A), if S/A is accepted, the shipper will receive a shipping order (S/O) to pick up empty containers from a depot and load the containers. Then, the carrier will offer a mates receipt (M/R) to the shipper. The shipper bears it to exchange bill of lading (B/L) and posts it to the consignee. The shipping agent at port of discharge informs the consignee to deliver the goods when it arrives. After the payment of all fees, the consignee uses B/L to exchange the delivery order (D/O) and takes delivery of goods.

As described above, a liner ship serves a large number of shippers; also, a shipper can choose one carrier from many competitors. In the procedure of decision making, the schedule and itinerary of the liner service will be firstly taken into account. Besides those, the freight is also another issue to be considered. There are some other factors affect the decision making such as the liner service level, the security of the area along the itinerary and so on. More importantly, the shippers are allowed to cancel the shipping contract signed with the carrier. Therefore, the actual amount of cargo shipment demand is highly uncertain. In practice, it is fluctuant. The slump of global seaborne trade caused by US subprime mortgage crisis is a convincing example. This view has been confirmed by the statistical data from Neptune Orient Lines (NOL), which is the global leader in container shipping. For the period November 15-December 26, Marcus30 reported that NOL carried 218,100 FEUs globally, equivalent to 24% fewer boxes than in the same period in 2007.

1.4. *Contributions*

The contributions of this paper are threefold: firstly, the assumption that the cargo shipment demand is deterministic in previous researches is not reasonable. This motivates us to seek a more reasonable assumption for our problem. In this paper, we assume that the cargo shipment demand between two ports on each liner trade route is a random variable following normal distribution with a given mean value and standard deviation. But this will create another issue: the liner shipping carrier has to satisfy the container transportation requirement on each liner trade route. Since the demand is assumed to be a random variable, how to formulate this constraint? If the fleet size and mix are deterministic, then its transportation capacity is deterministic; however, the demand is uncertain, this creates a risk that the fleet could not satisfy the container delivery requirement. In order to deal with this problem, we introduce the risk level concept and apply the chance constraints. The risk level refers to the possibility that the fleet's transportation capacity is insufficient. Then, we use the chance constraints to explore the possibility that the fleet could deliver the containers. Furthermore, we assume all unsatisfied containers are regarded lost. Dealing with the demand uncertainty is the main contribution of this paper. Moreover, we will study the sensitivity analysis of the effect of risk level on the fleet deployment plan.

Secondly, all of the previous researches either neglected or predetermined the cruising speeds of vessels in their proposed models. However, experiment data shows that the fuel consumption, the biggest part occupied in the total shipping cost of a vessel, is significantly affected by its cruising speed. It increases exponentially with the speed increasing one knot or more.[13] Thus, the cruising speed has an important impact on the shipping cost of a vessel. Within the consideration of its effect on the optimal fleet planning, it is regarded as a decision variable in this paper to seek the optimal cruising speed.

Thirdly, most researches considered the planning of fleet's size, mix and deployment separately; however, Agarwal and Ergun[31] addressed that they inter-affect on each other. In this paper, we extended and improved the models for LSFP built in previous researches by integrating the fleet size, mix and deployment in the decision-making procedure of the optimal fleet planning. In the procedure of decision-making, the carrier can use its own or chartered ships from other carriers to carry containers, even can charter-out some ones to other carriers.

This paper is organized as follows: Section 2 describes the liner ship fleet planning problem and introduces the notations and assumptions in this paper. Section 3 develops a mixed-integer nonlinear programming model with chance constraints to deal with the cargo shipment demand uncertainty. Section 4 explores the linearity relationship between the fuel consumption and the cruising speed and converts the MINLP model into a MILP model. Section 5 randomly generates numerical examples to illustrate our model and studies the impact of risk level on fleet deployment. Section 6 summarizes and discusses our work.

2. Problem Description, Assumptions and Notations

Consider a liner container shipping company operating a set of routes with given liner service frequencies for a short-term planning horizon to pick up and delivery cargoes from shippers. Given a cargo shipment demand pattern within a short-term horizon, the liner container shipping company has to deal with the short-term LSFP problem: to decide which types of own ships and how many of each type are used/chartered out (if any), or which types of chartered ships and how many of each type are needed (namely decide the fleet size and mix); furthermore, how to assign those ships in the fleet to the liner trade routes (namely decide the fleet deployment) so that the planned fleet provides a regular liner service on each route to pick up and delivery cargoes from one port to another. The following will explain the terms appeared in this paper: itinerary, charter strategies and chance constraints.

2.1. *Itinerary*

In most liner services, ship schedules and itineraries are fixed and services are offered on a regular basis. A liner service itinerary is defined as a sequence of ports visited by ships. The direction from origin to destination is called outbound direction, contrarily, is called inbound direction. A 'voyage' is defined as one round-trip shown in Fig. 1.

The outbound direction and inbound direction may be asymmetric. As the Fig. 1 shows, Ningbo Port is not visited in the outbound direction, while

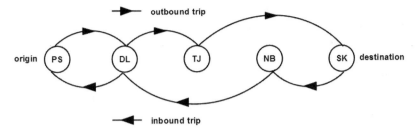

Fig. 1   An example of liner ships' itinerary.

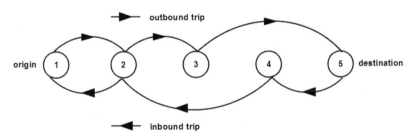

Fig. 2   The number code of liner ships' itinerary.

Tianjin Port is not visited in the inbound direction. The sequence of ports in this itinerary is: Pusan–Dalian–Tianjin–Shekou–Ningbo–Dalian–Pusan. In order to simplify the expression of ports sequence, we code the ports by numbers. Due to there are totally five ports in the itinerary, thus, number 1, 2, 3, 4 and 5 are used to present the ports. The smallest and the largest number present the origin and the destination port respectively. By using the number code, the sequence of ports is expressed as: 1 — 2 — 3 — 5 — 4 — 2 — 1 (see Fig. 2).

It is noted that in our work, we assume all routes have specific origin and destination port. This case may not be common in practice because some liner shipping company does not follow this convention. Sometimes, the origin and destination port are variable. However, in order to simplify our problem, we insist to make such assumption and moreover, all ships are empty before they start their journey.

2.2. *Charter strategies*

There are three main chartering ways: voyage charter, time charter, and bareboat charter.

A voyage charter, which is the most popular way in chartering market, provides transport for a specific cargo from one port to the other for a fixed price per ton. The ship owner has to complete the terms in charter-party and pay all fixed and variable costs. The charterer pays for the freight according to the charter-party. If completed over the due date of lay time (the period of time that a ship is available for loading and discharging of cargoes at a port) allowed in the charter-party, the owner will submit a claim of demurrage to the charterer. Conversely, if completed ahead, the charterer will submit a claim of dispatch to the owner.

A time charter gives the charterer operational control of the ship carrying his cargo, while leaving ownership and management of the vessel in the hands of the ship owner. The length of the charter is often a period of months or years (period charter). During this period, the ship owner continues to bear the all fixed costs except the capital repayment, tax and depreciation, but the charterer directs the commercial operations of the vessel and pays all variable costs.

Finally, if a charterer wishes to have full operational control of the ship, but does not wish to own it, a bareboat charter is arranged. The charterer manages the vessel and pays all costs except the capital repayment, tax and depreciation. In other words, the owner does not bear any cost except collecting the rent from the charterer. Due to the owner does not feel relieved to let the charterer operate his vessel and the management of a vessel for the charterer is not easy, bareboat charter is less popular compared with the above two ways though it is simplest. However, in order to simplify the problem, we assume a carrier can only adopt the bareboat charter in this paper if necessary.

Due to buying ships is a much huger capital investment compared with chartered ships, it is excluded in the short-term LSFP. According to the latest news released by China Shipping (Group) Company, it will book 8 Post-Suezmax ships with the shipping capacity of 12,600 TEUs from Samsung Engineering Company. The cost of a ship is about $170 million, and the total trade will exceed $1.3 billion.[32] However, the annual rent fee of a ship is only $22 million.[33] Thus, for the short-term fleet planning, chartering ships is preferred than buying ships.

2.3. *Chance constraints*

As aforementioned, the cargo (container) shipment demand between any two ports on each liner service route is assumed following a normal distribution. This assumption may lead to another problem: since the demands are uncertain, one can hardly find any decision which would definitely exclude later constraint violation caused by unexpected random effects, in other words, once the decisions in LSFP problem are determined, the fleet of ships may be unable to fully meet the pickups and deliveries requirement for its customers, even though the expected demands along the route do not exceed the fleet capacity. Once such case happens, it implies losing money for this liner container shipping company. Since it is hardly unavoidable, the liner container shipping company hopes that it happens at a low possibility as possible.

In order to reduce the possibility of the occurrence that the liner container shipping company cannot satisfy the customers' demand, such constraint violation can often be balanced afterwards by some compensating decisions which are considered as a penalization for constraint violation. However, the compensation cannot be modeled by cost in this paper because the cargo shipment demand is not realized. In such circumstances, we would rather insist on decisions guaranteeing feasibility 'as much as possible'. This loose term refers once more to the fact that constraint violation can almost never be avoided because of unexpected extreme events. On the other hand, when knowing or approximating the distribution of the random parameter, it makes sense to call decisions feasible whenever they are feasible with high probability, i.e., only a low percentage of realizations of the random parameter leads to constraint violation under this fixed decision. Therefore, we formulate the constraint that the liner container shipping company should satisfy the customers' demand as a probabilistic form in this paper, which is called chance constraint. The probability of the constraint violation is called a risk level in this study. It indicates that if the liner container shipping company makes a decision which satisfies the chance constraint, the event that the customers' demand cannot be met will occur at most with this probability. For those unmet cargoes, we regarded they are lost.

Here, we specially explain the term "frequency of liner service", which is the most important feature of liner shipping. It is defined as the time interval between two successive liner services.[21] For example, the shipping company Oriental Oversea Container Line (OOCL) provides a weekly liner service on the route Asian-Pacific Express.[2] In other word, the liner service frequency on that route is 7 days.

We also make another three assumptions: first, the cargo demand from ports to ports on a route are regarded to be independent random variables following normal distributions with given mean and standard deviation; second, the waiting time from chartering a ship to being serviced is ignored; thirdly, the number of available ships is finite; fourthly, we assume the ships are empty when they starts their journey.

2.4. *Notations*

Notations used in this paper are listed as follows:

Sets

K	Set of ship types' number code
M_r	Set of ports' number code on route r
R	Set of liner trade routes' number code
$\mathbb{Z}^+$	Set of positive integer
$\mathbb{R}^+$	Set of positive real number

Deterministic Parameters

c_{kr}:	Shipping cost of a ship of type k on route r (\$/voyage)
c_k^{in}:	Cost of chartering-in a ship of type k in the planning horizon (\$/ship)
c_k^{out}:	Profit of chartering-out ship of type k in the planning horizon (\$/ship)
d_r:	Length of route r (mile)
d_r^{ij}:	Distance from port i to port j on route r (mile)
e_k:	Lay-up cost for a ship of type k (\$/day)
M_r^d:	Number code of the destination port on route r
$N_k^{\max}$:	Maximum amount of the ships owned by the carrier of type k
$NCI_k^{\max}$:	Maximum amount of the ships of type k chartered-in from other carriers
S_{kr}^L:	Lower bound of the cruising speed of a ship of type k on route r (mile/hr)
S_{kr}^U:	Upper bound of the cruising speed of a ship of type k on route r (mile/hr)
T:	Length of the short-term planning horizon (day)
T_k:	Shipping season for a ship of type k in the planning horizon (day)
V_k:	Capacity of a ship of type k (TEUs)
α_r:	Risk level of the capacity of fleet is insufficient on route r

Random Data

ξ_r^{ij}:   Cargo demand from port i to port j on route r in the planning horizon (TEU)

μ_r^{ij}:   Mean value of ξ_r^{ij} (TEUs)

σ_r^{ij}:   Standard deviation of ξ_r^{ij} (TEUs)

η_r^{ij}:   Flow of containers on the segment from port i to port j on route r in the planning horizon (TEUs)

Decision variables

f_r: Frequency of liner service on route r (day/service)

n_{kr}:   Number of owned ships of type k sailing on route r

n_{kr}^{in}:   Number of chartered-in ships of type k sailing on route r

n_k^{out}:   Number of chartered-out ships of type k to other carriers

s_{kr}:   Cruising speed of ships of type k on route r

x_{kr}:   Number of voyages of ships of type k on route r

y_{kr}:   Lay-up days of ships of type k on route r

3. A Mixed Integer Nonlinear Programming Model with Chance Constraints

In this section, a MINLP model is developed to minimize the cost of shipping the containers in the short-term planning horizon.

The expenditure is consists of three components: shipping cost, lay-up cost and chartered rent. They are calculated as follows:

Expenditure in shipping costs:

$$C^s = \sum_{r \in R} \sum_{k \in K} c_{kr}(s_{kr}) x_{kr}$$

Expenditure in lay-up costs:

$$C^l = \sum_{r \in R} \sum_{k \in K} e_k y_{kr}$$

Expenditure in chartering-in ships:

$$C^{ci} = \sum_{r \in R} \sum_{k \in K} n_{kr}^{in} c_k^{in}$$

Thus, the total cost equals to:

$$C = \sum_{r \in R} \sum_{k \in K} (c_{kr}(s_{kr}) x_{kr} + e_k y_{kr} + n_{kr}^{in} c_k^{in}) \tag{1}$$

The $c_{kr}(s_{kr})$ in the above Eq. (1) denotes the shipping cost per voyage for a ship of type k on route r. It includes the port charges, canal fees, fuel cost, maintenance cost, insurance cost, administrative cost and crew cost, etc.[21] As explained in Stopford,[13] the cruising speed affects the fuel cost, thus, the shipping cost per voyage for a ship should be a function with respect to cruising speed. Based on the cost function formulated by Perakis and Jaramillo,[21] the shipping cost per voyage can be expressed as the following equation:

$$c_{kr}(s_{kr}) = \bar{\lambda}_{kr} s_{kr}^2 + \tilde{\lambda}_{kr}/s_{kr} + \hat{\lambda}_{kr} \tag{2}$$

where $\bar{\lambda}_{kr}, \tilde{\lambda}_{kr}$ and $\hat{\lambda}_{kr}$ are parameters.

Remark 1 There are some other different cost functions: Cullinane and Khanna,[34] Drewry,[35] Shintani *et al.*[36] and Takano and Arai.[37] These researches investigated the shipping cost by various ships' size. The regression analysis was applied based on the cost data. Thus, they formulated the shipping cost as a function with respective to the ship size. It is noted that the parameters in the regression functions of cost developed in those studies are dependent on the cost data space. Those regression models are valid only for a specific data space. In other words, those regression functions of shipping cost are not available in this paper; if employed, the parameters need to be calibrated.

Since we suppose liner operators can charter out some ships to other carriers to earn the profit, our objective function is formulated as follows:

$$[\mathbf{P}] \quad \min \sum_{r \in R} \sum_{k \in K} (c_{kr}(s_{kr}) x_{kr} + e_k y_{kr} + n_{kr}^{in} c_k^{in}) - \sum_{k \in K} n_k^{out} c_k^{out} \tag{3}$$

Subject to:

$$\Pr\left(\sum_{k \in K} x_{kr} V_k \geq \eta_r^{ij}\right) \geq 1 - \alpha_r, \quad i = 1, \ldots, M_r^d - 1;$$

$$j = i+1, \ldots, M_r^d; \ \forall r \in R \tag{4}$$

$$\Pr\left(\sum_{k \in K} x_{kr} V_k \geq \eta_r^{ij}\right) \geq 1 - \alpha_r, i = 2, \ldots, M_r^d;$$

$$j = 1, \ldots, i-1; \ \forall r \in R \tag{5}$$

$$\frac{d_r}{24s_{kr}} \times \frac{x_{kr}}{n_{kr} + n_{kr}^{in}} + \frac{y_{kr}}{n_{kr} + n_{kr}^{in}} \leq T, \quad \forall r \in R, \ k \in K \tag{6}$$

$$\frac{y_{kr}}{n_{kr} + n_{kr}^{in}} \geq T - T_k, \quad \forall r \in R, \ k \in K \tag{7}$$

$$\sum_{r \in R} n_{kr} \leq N_k^{\max}, \quad \forall k \in K \tag{8}$$

$$\sum_{r \in R} n_{kr}^{in} \leq NCI_k^{\max}, \quad \forall k \in K \tag{9}$$

$$n_k^{out} = N_k^{\max} - \sum_{r \in R} n_{kr}, \quad \forall k \in K \tag{10}$$

$$S_{kr}^L \leq s_{kr} \leq S_{kr}^U, \quad \forall k \in K, \ r \in R \tag{11}$$

$$x_{kr}, y_{kr}, n_{kr}, n_{kr}^{in}, n_k^{out} \in \mathbb{Z}^+ \cup \{0\}; \quad s_{kr} \in \mathbb{R}^+ \cup \{0\}, \ \forall k \in K, \ r \in R \tag{12}$$

Constraints (4) and (5) are chance constraints to investigate the possibilities that the fleet's capacity is sufficient on route r in the outbound trip and inbound trip respectively. When a ship sails from port i to port j, the containers on the ship includes those will be unloaded at port j from previously visited ports and those loaded at port i to the foregoing ports, it is referred to as a term flow of a segment from port i to port j in this paper. The containers loaded in the ship on any voyage segment can not exceed the capacity of the ship. Segment flow on each route is calculated as follows:

(a) *For outbound trip*

$$\eta_r^{ij} = \sum_{p=1}^{i} \sum_{q=j}^{M_r^d} \xi_r^{pq}, \quad i = 1, \ldots, M_r^d - 1; \ j = i+1, \ldots, M_r^d; \ \forall r \in R \tag{13}$$

(b) *For inbound trip*

$$\eta_r^{ij} = \sum_{p=i}^{M_r^d} \sum_{q=1}^{j} \xi_r^{pq}, \quad i = 2, \ldots, M_r^d; \ j = 1, \ldots, i-1; \ \forall r \in R \tag{14}$$

Remark 2 Due to the cargo demand from port i to port j on a route is assumed to follow inter-independent normal distribution, according to the probability theory, the flow of containers on the segment from port i to port j on a route also follows a normal distribution.

(a)$'$ *For outbound trip*

$$\eta_r^{ij} \sim N\left(\sum_{p=1}^{i}\sum_{q=j}^{M_r^d}\mu_r^{pq}, \sum_{p=1}^{i}\sum_{q=j}^{M_r^d}(\sigma_r^{pq})^2\right) \tag{15}$$

(b)$'$ *For inbound trip*

$$\eta_r^{ij} \sim N\left(\sum_{p=i}^{M_r^d}\sum_{q=1}^{j}\mu_r^{pq}, \sum_{p=i}^{M_r^d}\sum_{q=1}^{j}(\sigma_r^{pq})^2\right) \tag{16}$$

Thus, the above constraints (4) and (5) can be respectively rewritten as follows:

$$\sum_{k \in K} x_{kr} V_k \geq \Phi^{-1}(1-\alpha_r)\sqrt{\sum_{p=1}^{i}\sum_{q=j}^{M_r^d}(\sigma_r^{pq})^2} + \sum_{p=1}^{i}\sum_{q=j}^{M_r^d}\mu_r^{pq},$$

$$i = 1,\ldots,M_r^d - 1; \ \ j = i+1,\ldots,M_r^d; \ \ \forall r \in R \tag{17}$$

$$\sum_{k \in K} x_{kr} V_k \geq \Phi^{-1}(1-\alpha_r)\sqrt{\sum_{p=i}^{M_r^d}\sum_{q=1}^{j}(\sigma_r^{pq})^2} + \sum_{p=i}^{M_r^d}\sum_{q=1}^{j}\mu_r^{pq},$$

$$i = 2,\ldots,M_r^d; \ \ j = 1,\ldots,i-1; \ \ \forall r \in R \tag{18}$$

where $\Phi^{-1}(\bullet)$ is the inverse function with respective to probability $\bullet$.

Constraint (6) ensures the time of a ship sailing on sea and its lay-up time should not exceed the planning horizon. Rewrite it as follows:

$$\frac{d_r x_{kr}}{24 s_{kr}} + y_{kr} \leq T(n_{kr} + n_{kr}^{in}), \quad \forall r \in R, \quad k \in K \tag{19}$$

where the length of route r, d_r, can be calculated by the following formulation:

$$d_r = \sum_{i,j \in M_r} d_r^{ij}, \quad \forall r \in R \tag{20}$$

The least lay-up days for a ship of type k on route r are through constraint (7). We can rewrite is as follows:

$$y_{kr} \geq (T - T_k)(n_{kr} + n_{kr}^{in}), \quad \forall r \in R, \ k \in K \tag{21}$$

Constraints (8) and (9) ensure the number ships of own and chartered-in should not exceed the maximum available ships. Constraint (10) provides the calculation formulation of the number of chartering-out ships.

Constraint (11) provides the feasible bounds for cruising speed of a ship of type k on route r. Constraint (12) requires that all variables are nonnegative.

The frequency is the most important property in liner shipping, it could be obtained by the following formula:

$$f_r = \left[\frac{T}{\sum_{k \in K} x_{kr}} \right], \quad \forall r \in R \tag{22}$$

where $[\bullet]$ denotes the largest integer less than $\bullet$.

4. A Linearized Approach

According to one technique report 38, for ships in the range of up to 1,500 TEU, the speed is between 9 and 25 knots per hour, with the majority of ships sailing at some 15–19 knots. The most popular speed for the 1,500–2500 TEU ships is 18–21 knots. In the 2,500–4,000 TEU range, 90% of the ships have a speed of 20–24 knots. 71% of the 4,000–6,000 TEU ships have a speed of 23–25 knots. Finally, 80% of the ships that are larger than 6,000 TEU have speed of 24–26 knot. For the new generation of container ships we let the speed to be between 25–26 knots per hour. The design fuel consumption is assumed to be the upper bound of each interval. Table 1 shows the design fuel consumption of ships with different capacities from 1000 TEUs to 10000 TEUs.

Table 1 Design fuel consumption of ships with different capacities.

Capacity (TEU)	F^* (tons/day)	Capacity (TEU)	F^* (tons/day)
1000	40	4000	145
1250	50	5000	165
1500	60	6000	185
1750	70	7000	205
2000	80	8000	225
2250	90	9000	245
2500	100	10000	265

We use the data from Table 1 and plot the fuel consumption graph against the speed in the given interval. We can observe that the function can be nicely approximated by a linear function in the given interval. For example, for the ships with the capacities of 1,000–1,500 TEU and given F^* in Table 1, the fuel consumption against the speed is depicted in Fig. 3. Also, the absolute and relative errors of the approximated value with the real value against the speed are illustrated in the below Fig. 4 and Fig. 5.

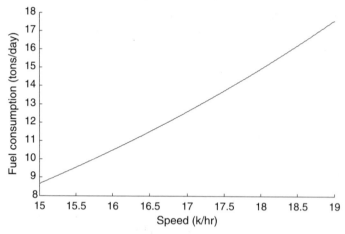

Fig. 3 Fuel consumption behaviors for the ships with capacities between 1,000 and 1,500 TEU.

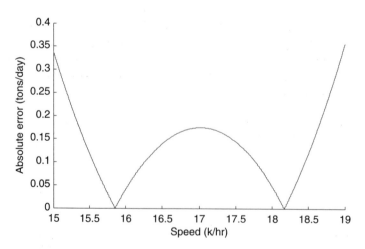

Fig. 4 Absolute error of approximated and real fuel consumption.

These two figures indicate that the maximum absolute error is less than 0.4 tons per day and the relative error are less than 0.04% compared with the real fuel consumption. For all range, they are also displayed in Fig. 6(a)∼(c). From Fig. 6(b), it indicates that the maximum error between the approximation and the real value of fuel consumption is less than 2 tons per day with the speed ranges from 20 knots to 24 knots per day. And

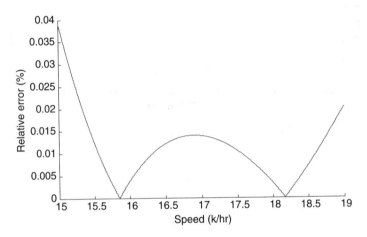

Fig. 5   Relative error of approximated and real fuel consumption.

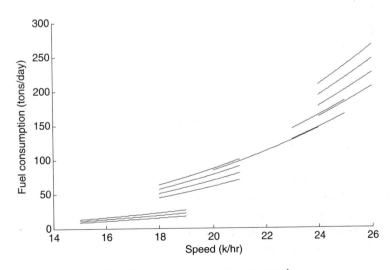

Fig. 6(a)   Fuel consumption vs. speed.

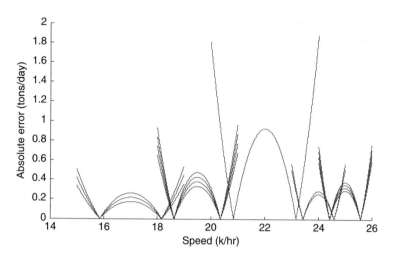

Fig. 6(b)   Absolute error of approximated and real fuel consumption.

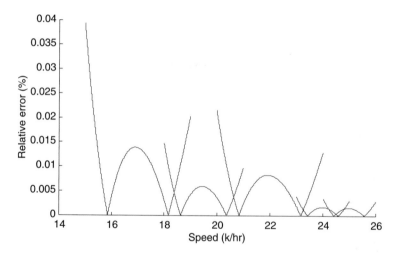

Fig. 6(c)   Relative error of approximated and real fuel consumption.

the maximum relative error shown in Fig. 6(c) is less than 0.04%. This implies that the approximation of the fuel consumption is acceptable and the performance is very well.

Therefore, the fuel consumption with respect to speed can be approximated by a linear function as follows:

$$F_{kr} = a'_k s_{kr} + b'_k \qquad (23)$$

Table 2   The slop and intercept of the approximated linear function.

Capacity (TEU)	a'_k	b'_k	Max absolute error (tons/day)	Max relative error (%)
1,000	2.2271	−25.0653	0.3563	0.0393
1,250	2.7838	−31.3317	0.4453	0.0393
1,500	3.3406	−37.5980	0.5344	0.0393
1,750	8.6357	−111.9076	0.6732	0.0148
2,000	9.8693	−127.8944	0.7694	0.0148
2,250	11.1030	−143.8812	0.8656	0.0148
2,500	12.3367	−159.8680	0.9618	0.0148
4,000	15.2610	−222.9008	1.8791	0.0216
5,000	18.2567	−291.7981	0.5108	0.0039
6,000	20.4696	−327.1676	0.5728	0.0039
7,000	21.8792	−364.2990	0.5875	0.0036
8,000	24.0138	−399.8403	0.6449	0.0036
9,000	26.1483	−435.3817	0.7022	0.0036
10,000	28.2829	−470.9231	0.7595	0.0036

where a'_k and b'_k are the slop and intercept and s_{kr} is the speed of a ship of type k on route r. They are listed in the below Table 2.

Now the shipping cost in Eq. (2) can be rewritten as follows:

$$c_{kr}(s_{kr}) = \tilde{\lambda}'_{kr}/s_{kr} + \hat{\lambda}'_{kr} \tag{24}$$

Let $z_{kr} = \frac{x_{kr}}{s_{kr}}$ ($\forall k, r$), and substitute it into constraint (11), then we have:

$$\frac{x_{kr}}{S^U_{kr}} \leq z_{kr} \leq \frac{x_{kr}}{S^L_{kr}}, \quad \forall k, r \tag{25}$$

Finally, the new MILP model with chance constraints is proposed as follows:

$$\min = \sum_{r \in R}\sum_{k \in K}(\tilde{\lambda}'_{kr}z_{kr} + \hat{\lambda}'_{kr}x_{kr} + e_k y_{kr} + n^{in}_{kr}c^{in}_k) - \sum_{k \in K} n^{out}_k c^{out}_k \tag{26}$$

s.t. constraints (4) to (12).

5. Numerical Example

In this section, a one-year planning for a fleet consisting of 5 types of ships operated on 8 liner trade routes by OOCL is studied: Central China Express (CCX), China Pakistan Express (CPX), Gulf Indian Subcontinent Strait Service (GIS), India US Express (IDX), North & Central China East Coast Express (NCE), New Zealand Express (NZX), South China East Coast

Table 3   Ports calling sequences in the liner trade routes.

Routes code	Port calling sequence and number code
1 (CCX)	Los Angeles/Oakland/Pusan/Dalian/Xingang/Qingdao/Ningbo/ Shanghai/Pusan/Los Angles (1-2-3-4-5-6-8-7-3-1)
2 (CPX)	Shanghai/Ningbo/Shekou/Singapore/Karachi/Mundra/Penang/ PortKelang/Singapore/Hong Kong/Shanghai (1-2-4-5-9-8-7-6-5-3-1)
3 (GIS)	Singapore/Port Kelang/Nhava Sheva/Karachi/Jebel Ali/Bandar Abbas/Jebel Ali/Mundra/Cochin/Singapore (1-2-4-6-7-8-7-5-3-1)
4 (IDX)	Colombo/Tuticorin/Cochin/Nhava heva/Mundra/Suez/Barchlona/ NewYork/Norfolk/Charleston/Barcelona/Suez/Colombo (1-2-3-4-5-6-7-8-9-10-7-6-1)
5 (NCE)	New York/Norfolk/Savannah/Panama/Pusan/Dalian/Xingang/ Qingdao/Ningbo/Shanghai/Panama/New York (1-2-3-4-5-6-7-8-9-10-4-1)
6 (NZX)	Singapore/Port Kelang/Brisbane/Auckland/Napier/Lyttelton/ Wellington/Brisbane/Singapore (1-2-3-4-5-7-6-3-1)
7 (SCE)	New York/Norfolk/Savannah/Panama/Kaohsiung/Shekou/Hong Kong/Panama/New York (1-2-3-4-5-6-7-4-1)
8 (UKX)	Southampton/Hull/Grangemouth/Southampton (1-2-3-1)

Source: OOCL website

Table 4   Types of ships with different capacities in the example.

Type	Capacity (TEU)	N_k^{max}	NCI_k^{max}
1 (ICE class)	2,808	1	10
2 (F class)	3,218	1	10
3 (P class)	4,500	9	25
4 (S class)	5,714	2	25
5 (SX class)	8,063	12	10

Note: the data in the first three columns is from OOCL website, the fourth column is hypothetical.

Express (SCE), UK Express (UKX); ICE class, F class, P class, S class and SX class of ships. The calling ports of each route and the associated number codes are listed in Table 3 and the types of ships are shown in Table 4. The data of costs are hypothetical due to the real data is not available.

Our discussion is twofold: first, 8 cases are tested to show the advantage of the model developed in this paper within consideration of the fleet size and mix, the cruising speed of ships and the liner service frequency simultaneously over those without. The comparison are shown in the Table 5; second, two cases are tested to discuss the effect of risk level on the fleet plan, shown in Fig. 7 and 8.

Table 5   Comparison of the 8 cases.

Case	Frequency	Fleet size and mix	Speed	Optimal solution	Difference with case 1	Proportion with case 1
1	×	×	×	415,867,758.4	0	100%
2	×	×	√	471,188,247.5	55,320,489.1	113.3%
3	×	√	×	461,005,592.1	45,137,833.7	110.8%
4	×	√	√	482,959,681.9	67,091,923.5	116.1%
5	√	×	×	489,881,584.6	74,013,826.2	117.8%
6	√	×	√	479,379,980	63,512,221.6	115.3%
7	√	√	×	489,887,214	74,019,455.6	117.8%
8	√	√	√	—	—	—

Note: decision variables are denoted by ×, otherwise, √.

From the Table 5, we can find that the Case 1 is the cheapest, in which the frequency, fleet and speed are taken into account. Compared Case 1 and Case 2, speed is decision variable in Case 1 while it is not in Case 2, the optimal solution to Case 2 is the feasible solution to Case 1, thus, the objective value of Case 2 is larger than that of Case 1. Similarly, if the fleet size and mix are also predetermined and fixed (Case 3 and Case 4), Case 4 has a larger optimal solution compared with Case 3. For Case 5 and Case 6, we can get the same conclusion if we set the same value of frequency in the two cases. If they are different, it is possible that Case 6 is more economical than Case 5. The similar conclusion can be obtained for Case 7 and Case 8 by the same deduction. It is noticed that there is no feasible solution to Case 8, in which the frequency, fleet size and mix and speed are all predetermined and fixed. This implies that whatever deployment of the fleet, it can not satisfy the demand requirement. If we decrease the demand, it may have feasible solution, depending on the cargo demand. In conclusion, any cases without taking the frequency, fleet size and mix and speed into account simultaneously are not the best fleet planning. The solution obtained by CPLEX of Case 1 is shown in Table 6.

In order to observe the role of α, the risk level, plays on the fleet plan, two examples with different transportation capacity of current fleet are generated and the corresponding results are shown in the following Fig. 7 and 8. These Figs. roughly indicate that with the decrease of α, the number of chartered-out ships has the decreasing trend, while the number of chartered-in ships has the increased trend. Compared the Figs., we found that Fig. 7(a~c) are more fluctuant than Fig. 8(a~c), this implies that α effects on the first example more than that on the second one. The transportation capacity of the current fleet in the first example is tighter

Table 6   Solution to case 1.

Ship types: 1 (ICE), 2 (F), 3 (P), 4 (S), 5 (SX)

Routes code	Fr	No. of voyages	Lay-up days	Speeds	Number of ships of each type on each route		No. of ships charter-out of each type
					Own	Charter-in	
1 (CPX)	3	74,-,-,-,47	45,-,-,-,60	15,-,-,-,15	-,-,-,-,-	3,-,-,-,2	-,-,1,-,10
2 (GIS)	5	51,21,-,-,-	30,15,-,-,-	18.21,15,-,-,-	-,1,-,-,-	2,-,-,-,-	
3 (IDX)	4	76,-,-,-,15	45,-,-,-,30	21.11,-,-,-,19.18	-,-,-,-,-	3,-,-,-,1	
4 (NCE)	9	19,-,-,-,21	15,-,-,-,30	17.86,-,-,-,20.62	1,-,-,-,1	-,-,-,-,-	
5 (SCE)	5	-,-,72,-,-	-,-,90,-,-	-,-,17.19,-,-	-,6,-,-	-,-,-,-,-	
6 (SIX)	16	-,2,-,-,20	-,15,-,-,60	-,15,-,-,16	-,-,-,1	-,1,-,-,1	
7 (SBX1)	5	-,-,61,11,-	-,-,75,20,-	-,-,17.93,16.4,-	-,-,-,1,-	-,5,-,-	
8 (SBX2)	5	10,-,53,9,-	15,-,90,20,-	24.35,-,21.51,22.23,-	-,2,1,-	1,-,4,-,-	

The number of own ships are used: 14
The number of charter-in ships is: 23
The number of charter-out ships is: 11
The objective value is: 415,867,758.4

Note: The elements of the five-dimensional vectors in the last six columns denote the correspond value to the ship of the five types; — denotes not available.

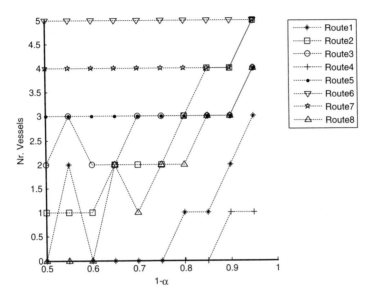

Fig. 7(a)   Deployment of chartered-in ships in Example 1 with tight capacity.

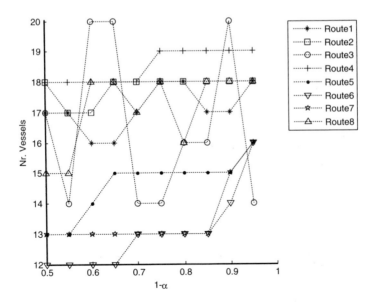

Fig. 7(b)   Deployment of own ships in Example 1 with tight capacity.

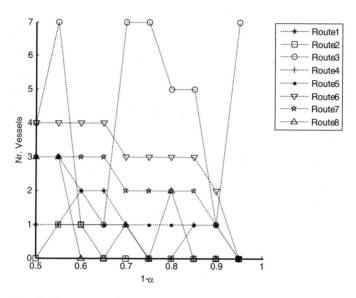

Fig. 7(c)   Deployment of chartered-out ships in Example 1 with tight capacity.

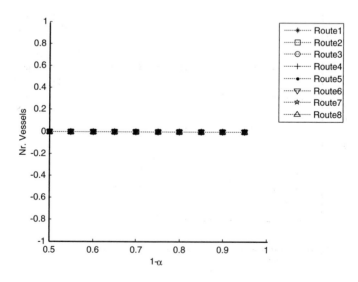

Fig. 8(a)   Deployment of chartered-in ships in Example 2 with loose capacity.

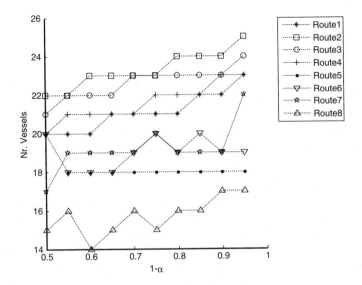

Fig. 8(b)   Deployment of own ships in Example 2 with loose capacity.

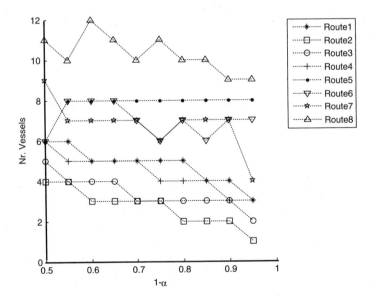

Fig. 8(c)   Deployment of chartered-out ships in Example 2 with loose capacity.

than that in the second example, especially for the number of chartered-in ships. (see Fig. 7(a) and Fig. 8(a)).

6. Summary and Conclusion

In this paper, we studied the liner fleet deployment problem within the cargo demand uncertainty. We first reviewed the previous related researches and addressed their limitations. The risk level is introduced to describe the probability that a liner operator could not satisfy the shippers' requirement. Then we built a mixed integer nonlinear programming model with chance constraints. Taken the fact that the fuel consumption cost for a ship can be approximated by a linear function with respect to its cruising speed, we proceed to build a mixed integer linear programming model that can approximate the mixed integer nonlinear programming model proposed in this study. Numerical examples were generated to show the application of this model and the CPLEX solver was employed to solve them. Lastly, the effect of risk level on the fleet deployment was studied. Generally we found that when the transportation capacity of the current fleet was tight, the fleet deployment was more sensitive to the risk level compared with that when the transportation capacity of the current fleet was loose. Though the GAMS solver can be applied into our model, it can not solve the model with large size. How to efficiently solve the problem is still worthwhile to be studied.

Acknowledgement

This paper was supported partially by the Neptune Orient Line Company Research Grant No. R-264-000-244-720.

References

1. UNCTAD, Review of Maritime Transportation, *Paper presented at the United Nations Conference on Trade and Development.* New York and Geneva, USA. http://www.unctad.org/en/docs/rmt2008_en.pdf (2008).
2. OOCL, *Sailing Schedule.* Available from: http://www.oocl.com/eng/ ourservices/eservices/sailingschedule/ (2009)
3. Ronen, D. (1983). Cargo ships routing and scheduling: Survey of models and problems, *Eur. J. Oper. Res.*, 12(2), pp. 119–126.
4. Ronen, D. (1993). Ship scheduling: The last decade, *Eur. J. Oper. Res.*, 71(3), pp. 325–333.
5. Perakis, A.N. (2002). Fleet operation optimization and fleet deployment. Costas Th.

6. Christiansen, M., Fagerholt, K. and Ronen, D. (2004). Ship routing and scheduling: status and perspectives, *Transport. Sci.*, 38(1), pp. 1–18.
7. Dantzig, G.B. and Fulkerson, D.R. (1954). Minimizing the number of tankers to meet a fixed schedule, *Nav Res. Logist. Q.*, 1, pp. 217–222.
8. Laderman, J., Gleiberman, L. and Egan, J.F. (1966). Vessel allocation by linear programming, *Nav Res. Logist. Q.*, pp. 315–320.
9. Everett, J.L., Hax., A.C., Lewinson, V.A. and Nudds, D. (1972). Optimization of a Fleet of Large Tankers and Bulkers: A linear Programming Approach, *Marine Tech.*, October, pp. 430–438.
10. Benford, H. (1981). A simple approach to fleet deployment, *Maritime Pol. Mng.*, 8(4), pp. 223–228.
11. Perakis, A.N. (1985). A second look at fleet deployment, *Maritime Pol. Mng.*, 12(3), pp. 209–214.
12. Lawrence, S.A. (1972). *International Sea Transport: The Years Ahead.* Lexington Books, 1st ed. Lexington, MA.
13. Stopford, M. (1997). *Maritime Eco.* 2nd ed. London: Routledge.
14. Lane, D.E., Heaver, T.D. and Uyeno, D. (1987). Planning and scheduling for efficiency in liner shipping, *Maritime Pol. Mng.*, 12(3), pp. 109–125.
15. Fagerholt, K. (1999). Optimal fleet design in a ship routing problem, *International Transactions in Oper. Res.*, 6(5), pp. 453–464.
16. Hsu, C.I. and Hsieh, Y.P. (2007) Routing, ship size and sailing frequency decision-making for a maritime hub-and-spoke container network, *Math Comput Model,* 45, pp. 89–916.
17. Brandao, J. (2009). A deterministic tabu search algorithm for the fleet size and mix vehicle routing problem, *Eur. J. Oper. Res.*, 195(3), pp. 716–728.
18. Perakis, A.N. and Papadakis, N. (1987). Fleet deployment optimization models Part 1, *Maritime Pol. Mng.*, 14(2), pp. 127–144.
19. Perakis, A.N. and Papadakis, N. (1987). Fleet deployment optimization models Part 2, *Maritime Pol. Mng.*, 14(2), pp. 145–155.
20. Papadakis, N. and Perakis, A.N. (1989). A Nonlinear Approach to the Multiorigin, Multidestination Fleet Deployment Problem, *Nav Res. Log.*, 36, pp. 515–528.
21. Perakis, A.N. and Jarammillo, D.I. (1991). Fleet deployment optimization for liner shipping Part 1. Background, problem formulation and solution approaches, *Maritime Pol. Mng.*, 18(3), pp. 183–200.
22. Jarammillo, D.I. and Perakis, A.N. (1991). Fleet deployment optimization for liner shipping Part 2. Implementation and results, *Maritime Pol. Mng.*, 18(3), pp. 235–262.
23. Powell, B.J. and Perakis, A.N. (1997). Fleet deployment optimization for liner shipping: an integer programming model, *Maritime Pol. Mng.*, 24(2), pp. 183–192.
24. Millar, H.H. and Gunn, E.A. (1991). Dispatching a fishing trawler fleet in the Canadian R. Agarwal, O., Ergun, , 2008. Ship Scheduling and Network Design for Cargo Routing in Liner Shipping, *Transport. Sci*, 42(2), pp. 175–196.

25. Vukadinović, K. and Teodorović, D. (1994). A fuzzy approach to the vessel dispatching problem, *Eur. J. Oper. Res.*, 76(1), pp. 155–164.

26. Cho, S.C. and Perakis, A.N. (1996). Optimal liner fleet routing strategies, *Maritime Pol. Mng.*, 23(3), pp. 249–259.

27. Mourão, M.C., Pato, M.V. and Paixão, A.C. (2001). Ship assignment with hub and spoke constraints, *Maritime Pol. Mng.*, 29(2), pp. 135–150.

28. Nicholson, T.A.J. and Pullen, R.D. (1971). Dynamic Programming Applied to Ship Fleet Management, *Oper. Res. Quart*, 22(3), pp. 211–220.

29. Xie, X.L., Wang, T.F. and Chen, D.S. (2000). A dynamic model and algorithm for fleet planning, *Maritime Pol. Mng.*, 27(1), pp. 53–63.

30. Marcus, H. (2009). NOL sees box volumes crash 24% in weeks. Available from: http://www.lloydslist.com/ll/news/viewArticle.htm?articleId=20017612522.

31. Agarwal, R. and Ergun, O. (2008). Ship schedule and network design for cargo routing in liner shipping, *Transport. Sci*, 42(2), pp. 175–196.

32. COSCO, released magnitude expansion treatment. Available from: http://www.shippingchina.com/static/brmaininfo/200708/32899.shtml. (2007).

33. EMC, *Evergreen Marine Corporation Negotiate Renting 126000 TEU Ships*. Available from: http://news.gcdcs.com/article/sort015/info-10505.html. (2007)

34. Cullinane, K. and Khanna, M. (1999). Economics of scale in large container ships, *J Transp. Econ Policy*, 33(2), pp. 185–208.

35. Drewry (2001). Ship Operating Costs Annual Review and Forecast. Drewry Shipping Consultants.

36. Shintani, K., Imai, A., Nishimura, E. and Papadimitriou, S. (2007). The container shipping network design problem with empty container repositioning, *Transport Res E-LOG*, 43(1), pp. 39–59.

37. Takano, K. and Arai, M. (2008). A genetic algorithm for the hub-and-spoke problem applied to containerized cargo transport, *J Maritime Sci Tech*, (in press, 2008).

38. MAN B&W Diesel A/S, Propulsion Trends in Container Vessels. Available from: http://www.manbw.com/files/news/filesof4672/P9028.pdf. Copenhagen, Denmark (2004).

CHAPTER 12

SHIP EMISSIONS, COSTS AND THEIR TRADEOFFS

Harilaos N. Psaraftis and Christos A. Kontovas

Laboratory for Maritime Transport,
National Technical University of Athens,
9, Iroon Polytechneiou Str.,
GR-157-73 Athens, Greece
hnpsar@mail.ntua.gr kontovas@mail.ntua.gr

Emissions from commercial shipping are currently the subject of intense scrutiny. Various analyses of many aspects of the problem have been and are being carried out and a spectrum of measures to reduce emissions is being contemplated. However, such measures may have important side-effects as regards the logistical supply chain, and vice-versa. Industry circles have also voiced the concern that low-sulphur fuel in SECAs (the so-called 'sulphur emissions control areas') may make maritime transport (and in particular short-sea shipping) more expensive and induce shippers to use land-based alternatives. A reverse shift of cargo from sea to land might ultimately increase the overall level of CO_2 emissions along the intermodal chain. This paper takes a look at various tradeoffs and may impact the cost-effectiveness of the logistical supply chain and present models that can be used to evaluate these tradeoffs. One of the key results is that speed reduction will always result in a lower fuel bill and lower emissions, even if the number of ships is increased to meet demand throughput. Another result is that cleaner fuel at SECAs may result in a reverse cargo shift from sea to land that has the potential to produce more emissions on land than those saved at sea. Various examples are presented.

1. Introduction

Air pollution from ships is currently at the center stage of discussion by the world shipping community and environmental organizations. The Kyoto protocol to the United Nations Framework Convention on Climate Change -UNFCCC (1997) stipulates concrete measures to reduce CO_2 emissions in order to curb the projected growth of greenhouse gases (GHG) worldwide.

Although some regulation exists for non-GHGs, such as SO_2, NO_x and others, shipping has thus far escaped being included in the Kyoto global emissions reduction target for CO_2 and other GHGs (such as CH_4 and N_2O). Even so, it is clear that the time of GHG non-regulation is rapidly approaching its end, and measures to curb future CO_2 and other GHG growth are being sought with a high sense of urgency and are very high on the agenda of the International Maritime Organization (IMO) and of many individual coastal states. In the forthcoming UNFCCC, which will take place in Copenhagen in December of 2009, shipping is expected to be included in the discussions on future GHG reduction. In that sense, various analyses of many aspects of the problem have been and are being carried out and a broad spectrum of measures is being contemplated. These measures can be considered to fall into three general categories: technical, market-based and operational.

Technical measures include more efficient ship hulls, energy-saving engines, more efficient propulsion, use of alternative fuels such as fuel cells, biofuels or others, "cold ironing" in ports (providing electrical supply to ships from shore sources), devices to trap exhaust emissions (such as scrubbers), and others, even including the use of sails to reduce power requirements. Market-based instruments (MBIs) are classified into two main categories, Emissions Trading Schemes (ETS) and Carbon Levy schemes (also known as International Fund schemes). Finally, operational schemes mainly involve speed optimization, optimized routing, improved fleet planning, and other, logistics-based measures.

Some of these measures, important in their own right as regards emissions reduction, may have non-trivial side-effects as regards the logistical supply chain. For instance, measures such as (a) reduction of speed, (b) change of number of ships in the fleet, (c) possibly others, will generally entail changes (positive or negative) in overall emissions, but also in other logistics and cost-effectiveness attributes such as in-transit inventory and other costs. Also, industry circles have voiced the concern that the mandated use of lower-sulphur fuel in some regions or globally may make maritime transport (and in particular short-sea shipping) more expensive and induce shippers to use land-based alternatives (mainly road). A reverse shift of cargo from sea to land would go against the drive to shift traffic from land to sea to reduce congestion, and might ultimately increase the overall level of CO_2 emissions along the intermodal chain. In that regard, in Europe one can already see a potential conflict between two policies: (a) the designation of certain areas as "sulphur emissions controlled areas" (or SECAs), such as the Baltic Sea, the North Sea and the English Channel, and (b) the stated Transport Policy goal of shifting cargo off the roads and onto ships and railways.

Typical problems in the maritime logistics area include one or a combination of problems from the following generic list (which is non-exhaustive):

- Optimal ship speed
- Optimal ship size
- Routing and scheduling
- Fleet deployment
- Fleet size and mix
- Weather routing
- Intermodal network design
- Modal split
- Transshipment
- Queuing at ports
- Terminal management
- Berth allocation
- Supply chain management

The traditional analysis of these problems is in terms of cost- benefit criteria from the point of view of the logistics, operator, shipper, or other end-user. Such analysis typically ignores environmental issues. Green maritime logistics tries to bring the environmental dimension into the problem, and specifically the dimension of emissions reduction, by trying to analyze the tradeoffs that are at stake and exploring win-win solutions.

It is also important to realize that two different settings can be analyzed, the strategic setting and the operational one. The distinction between the two is important, and one that is not mentioned frequently. Let us clarify the difference between the two by an example.

A spokesman from Germanischer Lloyd (GL) has been recently quoted as follows: "We recommend that ship-owners consider installing less powerful engines in their newbuildings and to operate those container vessels at slower speeds," (Lloyds List, 2008a). By 'slower speeds' it is understood that the current regime of 24–26 knots would be reduced to something like 21–22 knots. But some trades may go as low as 15–18 knots, according to a 2006 study by Lloyds Register (Lloyds List, 2008b). An obvious reason for suggesting such speed reduction is twofold: fuel costs and emissions.

Implementing the aforementioned speed reduction would only make sense in a strategic setting, by modifying the design of the ship, including hull shape, by installing smaller engines in future newbuildings, by modifying the propeller design, etc. In such a setting however, one would have to also investigate not only differences in emissions produced by these modified lower-speed designs, but also other possible ramifications. These may include emissions differentials by the shipyards that produce these ships, as well as any difference in emissions when these ships would be recycled. This strategic approach to the emissions problem is also known as the 'life-cycle' approach. It is an important component in the quest to formulate possible strategic decisions and policies to curb emissions from shipping in the long run.

It is not the scope of this paper to examine all of the problems identified above from an environmental perspective. That will take years to accomplish. Rather, the limited number of models examined in this paper primarily focus on operational scenarios and mainly serve to highlight some of the trade-offs that are at stake in these scenarios, so as to motivate further work in this area.

The rest of this paper is organized as follows. Section 2 reports on relevant background. Section 3 describes some basics on emissions. Section 4 describes a simple logistical scenario to investigate the effects of speed reduction. Section 5 introduces the concept of the cost to avert a tonne of CO_2 and Section 6 examines the issue of port time in the quest to reduce emissions. Section 7 examines the effect of speed reduction at SECAs and Section 8 looks into possible side-effects of cleaner fuels on modal split. Finally Section 9 presents the paper's conclusions.

2. Background

We start by stating that even though the literature on the broad area of ship emissions is immense, the literature on the specific topic (link between emissions and maritime logistics) is scant. There are a number of papers that consider the economic impact of speed reduction especially for container vessels. Andersson (2008) considered the case of a container line where the speed for each ship reduced from 26 knots to 23 knots and one more ship was added to maintain the same throughput. Total costs per container were reduced by nearly 28 per cent. Eefsen (2008) considered the economic impact of speed reduction of containerships and included the inventory cost. Cerup-Simonsen (2008) developed a simplified cost model to demonstrate how an existing ship could reduce its fuel consumption by a speed reduction in low and high markets to maximize profits. Corbett *et al.* (2009) applied fundamental equations relating speed, energy consumption, and the total cost to evaluate the impact of speed reduction. The paper also explored the relationship between fuel price and the optimal speed.

The situation is similar at the policy level: many activities, but little or nothing relating to the interface between emissions and logistics. Looking at developments at the IMO (International Maritime Organization) level, thus far progress as regards air pollution from ships has been mixed and rather slow. On the positive side, in November 2008 the Marine Environment Protection Committee (MEPC) of the IMO unanimously adopted amendments to the MARPOL Annex VI regulations. The main changes will see a progressive reduction in sulphur oxide (SOx) emissions from ships, with the global sulphur cap reduced initially to 3.50%, effective 1 January 2012; then progressively to 0.50%, effective 1 January 2020 (IMO, 2008a).

Furthermore, the report of Phase 1 of the update the 2000 IMO GHG Study (IMO, 2000) was presented, which was conducted by an international consortium led by Marintek, Norway (Buhaug, *et al.*, 2008). According to this study, total CO_2 emissions from shipping (both domestic and international) are estimated to range from 854 to 1,224 million tons (2007), with a 'consensus estimate' set at 1,019 million tons, or 3.3% of global CO_2 emissions. By comparison, electricity and heat production accounts for 35% of global CO_2 emissions, manufacturing industries and construction 18.2%, and transport (all modes) 21.7%. Among transport modes, road accounts for 51% of all CO_2 emissions, shipping (including fishing) for 25%, aviation for 20%, and rail for 4%. However, in terms of energy use and emissions per tonne-km, shipping ranks as the most environment-friendly transport mode, as can be seen in the following table.

Among ship types, according to the results of Phase 1, the three top fuel consuming categories of ships (and thus, those that produce most of the CO_2 emissions) are (i) container vessels of 3,000–5,000 TEUs, (ii) container vessels of 5,000–8,000 TEUs and (iii) RoPax Ferries with cruising speed of less than 25 knots. The common denominator of these three categories, which results in a high level of CO2 emissions, is their high speed, at least as compared to other ship types.

These findings are in line with those of Psaraftis and Kontovas (2008, 2009a). According to their analysis, containerships are the top CO_2

Table 1　Energy efficiency and emissions to the atmosphere (by mode).

Energy use	PS-Type container vessel (11,000 TEU)	S-Type container vessel (6,600 TEU)	Rail — electric	Rail — diesel	Heavy truck	Boeing 747-400
kWh/tkm	0.014	0.018	0.043	0.067	0.18	2.00

Emissions (g/tkm)	PS-Type container vessel (11,000 TEU)	S-Type container vessel (6,600 TEU)	Rail — electric	Rail — diesel	Heavy truck	Boeing 747-400
Carbon dioxide (CO_2)	7,48	8.36	18	17	50	552
Sulphur oxides (SO_X)	0.19	0.21	0.44	0.35	0.31	5.69
Nitrogen oxides (NO)	0.12	0.162	0.10	0.00005	0.00006	0.17
Particulate matters (PM)	0.008	0.009	n/a	0.008	0.005	n/a

Re. emissions for rail; the complete value chain for el-production is considered.
Source: Network for Transport and the Environment (Sweden).

emissions producer in the world fleet (2007, Lloyds-Fairplay database). Just the top tier category of container vessels (those of 4,400 TEU and above) are seen to produce CO_2 emissions comparable on an absolute scale to that produced by the entire crude oil tanker fleet (in fact, the emissions of that top tier alone are slightly higher than those of all crude oil tankers combined- see Fig. 1 above).

At the latest meeting of IMO's Marine Environment Protection Committee in London last July (MEPC 59) there continued to be a clear split between industrialized member states, such as Japan, Denmark and other Northern European countries, and a group of developing countries including China, India and Brazil, on how to proceed. The latter countries spoke in favor of the principle of "Common but differentiated responsibility" (CBDR) under the UNFCCC. In their view, any mandatory regime aiming

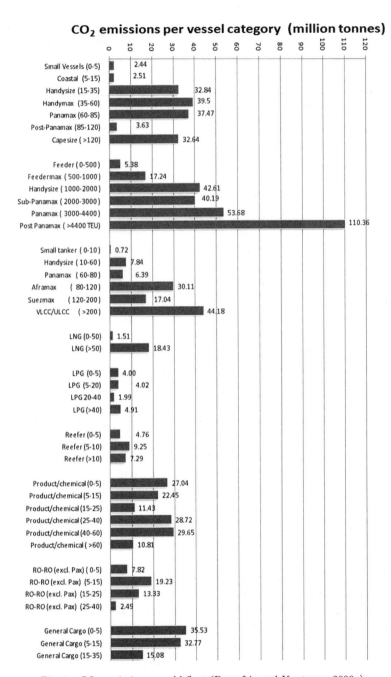

Fig. 1 CO_2 emissions, world fleet (Psaraftis and Kontovas, 2009a).

to reduce GHG emissions from ships engaged in international trade should be applicable exclusively to the countries listed in Annex I to the UNFCCC, therefore their strong wish is not to be included in any mandatory set of measures.

Due to 'political' reasons such as above, progress as regards regulating CO_2 and other GHGs continues to be very slow. In fact, the stated objective to finalize a mandatory Energy Efficiency Design Index (EEDI) of the environmental performance of new ships has not been reached yet. The same is true for the Energy Efficiency Operational Indicator (EEOI), which will be applicable to all ships. As a result, the IMO will not be in a position to have reached a clear position on these two indices in time for the United Nations Framework Conference for Climate Change (UNFCCC) that will be held in Copenhagen in December of this year, when a new climate agreement is expected to be reached, after Kyoto in 1997.

Without going into technical details regarding these two indices, one can state that the first index (EEDI) concerns the design of new ships and the second (EEOI) concerns the operation of all ships, new and existing. Both indices are ratios, in which the numerator is a complex function of all energy consumed by the ship, and the denominator includes a product of the ship's deadweight (or payload) and the ship's operational speed. The fact that speed is in the denominator means that the slower the ship goes, the higher both these indices will be, therefore the higher the ship will be ranked in terms of energy efficiency, both for design and for operation. No doubt about it, faster ships will score low as regards these indices.

The implication of this is unknown, other than the fact than in any ranking based on these indices, fast ships will have an unfavorable environmental performance vis-à-vis slower ships of the same capacity. In spite of extensive discussions on this topic, it is still not clear exactly how these indices will be used in future IMO rulemaking. In fact, these indices still have not been finalized, as certain issues still demand discussion and agreement.

Progress as far as other measures to regulate GHG emissions, such as MBIs has been even slower. Reaction to this concept has been even more pronounced, and it is not clear which among two main schemes, the Emissions Trading Scheme (ETS) and the Carbon Levy, will be eventually adopted. Certainly no agreement will be reached before the Copenhagen UNFCCC conference, and the latest IMO timetable on this issue goes into 2012.

What does slow progress on GHGs mean? And what if no agreement is reached at the IMO any time soon? This will certainly increase the pressure for regional approaches. In fact the European Commission is following IMO developments very closely, and has stated very clearly its intention to act alone if IMO's procedures take longer than previously anticipated. As regards GHGs, the anticipated approach of the Commission is to formulate an ETS, similar to that used in other land-based industries. The Commission has started the procedure for including air transport into its ETS scheme, and many think it will eventually do the same for shipping. Many ship owners' circles have voiced strong concerns that such a scheme would be complicated and unworkable.

Currently, European legislation mainly concerns the sulphur content of marine fuels. The maximum sulphur content for marine fuels according to EU directive 2005/33/EC is in line with MARPOL Annex VI. The implementation dates are differently from those agreed by the IMO under MARPOL Annex VI, but the main point is that currently all vessels sailing in the designated areas (SECAs) should use marine fuels with a maximum of 1.5% by mass content of sulphur. What is different from MARPOL is that the EU Directive sets a limit for all passenger vessels operating on regular service to or from EU ports to a maximum sulphur content of 1.5% (the same as in SECAs). This limit came into effect on August 11th, 2006 (EU directive 2005/33/EC, Article 4a). Furthermore, according to Article 4b of the same Directive, from January 1st, 2010 a 0.1% limit comes into effect for inland waterway vessels and ships at berth in EU ports with some exemptions.

Perhaps more interesting are developments on the logistics side: the European Commission states in their Freight Transport Logistics Action Plan launched in October 2007 that *"Logistics policy needs to be pursued at all levels of governance"*, which is also the reason behind this action plan as one in a series of policy initiatives to improve the efficiency and sustainability of freight transport in Europe. In the Freight Transport Logistics Action Plan a number of short — to medium-term actions is presented that will help Europe address its current and future challenges and ensure a competitive and sustainable freight transport system in Europe. Among the actions are the *"Green* transport corridors for freight". The Green Corridors are characterized by a concentration of freight traffic between major hubs and by relatively long distances of transport. Green Corridors should in all ways be environmentally friendly, safe and efficient. This is perhaps one of the few EU policy initiatives that aim to establish a clear connection between environment and logistics, even though this activity is still very much at its infancy. It is clear that the maritime mode will be involved in some of these Green Corridors, particularly those involving the Trans European Transport Networks (TEN-T's) and the Motorways of the Sea, and the question is, what ships, what types, what sizes, what speeds, how will they be utilized, and how will tradeoffs will be assessed.

In the United States, the Environmental Protection Agency (EPA) has established a tier-based timeline for implementing NOx emission standards to marine diesel engines that became effective in 2007. These standards are similar to those described in MARPOL Annex VI which has been ratified by the US in October 2008 although the Convention entered into force in May 2005. Canada has not yet ratified Annex VI, however Canada and the United States jointly proposed the designation of an Emissions Control Area (ECA) for specified portions of the US and Canadian coastal waters covering a total o 200 nm. At MEPC 59, the proposal was agreed in principle and will be voted during MEPC 60, scheduled for March 2010. If approved the ECA would enter into force in 2012.

On a local government basis, the State of California which is the home of the two busiest ports in the US has created a special agency, the California Air Resources Board (CARB) which is the primary source for ship emission regulations in California. On July 2008, CARB adopted the regulation "Fuel Sulfur and Other Operation Requirements for Ocean-Going Vessels within California Waters and 24 Nautical Miles of the California Baseline" that sets specific limits on the sulfur content of fuel used within 24 nm of the Californian coast.

In addition, the two busiest ports in the US (Long Beach and Los Angeles) both located in Southern California have introduced a series of voluntary incentive-based programs. On March 2008, the Board of Commissioners of the ports of Los Angeles (POLA) and Long Beach (POLB) authorized the Low-Sulphur Vessel Main Engine Fuel Incentive program to encourage operators to use cleaner fuels within 40 nm or 20 nm from Point Fermin. The program will pay the operators that will agree to use fuels that contain less than 0.2% sulphur the price difference between that fuel and IFO 380. Furthermore, the two ports offer a 15% discount on dockage fees to vessels that voluntary comply with the SPBP-OGV1 Vessel Speed Reduction Program and reduce their speed to 12 knots within 20 nm of Point Fermin while entering or leaving the ports.

3. Some Basics: Algebra of Emissions and Fuel Cost

Before logistical scenarios are examined, some basics have to be established first. Two are the main attributes of any logistical scenario that is viewed from a green perspective: the amount of emissions produced, and the cost. To calculate CO_2 emissions, one has to multiply bunker consumption by an appropriate emissions factor, F_{CO2}. The factor of 3.17 has been the empirical mean value most commonly used in CO_2 emissions calculations based on fuel consumption (see EMEP/CORINAIR (2002) and Endresen (2007)). According to the IMO GHG study (IMO, 2000), the actual value of this coefficient may range from 3.159 (low value) to 3.175 (high value). The update of the IMO 2000 study (Buhaug *et al.*, 2008), uses slightly lower coefficients, different for Heavy Fuel Oil and for Marine Diesel Oil. The actual values are 3.082 for Marine Diesel and Marine Gas Oils (MDO/MGO) and 3.021 for Heavy Fuel Oils (HFO). According to the report of the Working Group on Greenhouse Gas Emissions from Ships (IMO, 2008b), the group agreed that the Carbon to CO_2 conversion factors used by the IMO should correspond to the factors used by IPCC (2006 IPCC Guidelines) in order to ensure harmonization of the emissions factor used by parties under the UNFCCC and the Kyoto Protocol. In this paper we shall use the original value of 3.17 also used in Psaraftis and Kontovas (2008, 2009a) except for the example in Section 6 where the value of 3.13 has been used, noting that our emissions results will have to be scaled down by up to 5% if a lower emissions factor is used. Table 2 summarizes various emissions factors.

As regards SO_2, this type of emissions depends on the type of fuel used. One has to multiply total bunker consumption (in tonnes per day) by the percentage of sulphur present in the fuel (for instance, 4%, 1.5%, 0.5%,

Table 2 Comparison of emission factors kg CO_2/kg fuel. (IMO, 2008b).

Fuel type	GHG-WG 1/3/1	IPCC 2006 Guidelines			Revised 1996 guidelines
		Default	Lower	Upper	
Marine diesel and marine gas oils (MDO/MGO)	3.082	3.19	3.01	3.24	3.212
Low sulphur fuel oils (LSFO)	3.075	3.13	3.00	3.29	
High sulphur fuel oils (HSFO)	3.021				

or other) and subsequently by a factor of 0.02 to compute SO_2 emissions (in tonnes per day). The factor of 0.02 is exact, and is derived from the chemical reaction of sulphur with oxygen.

Finally, NO_x emissions depend on engine type. The ratio of NO_x emissions to fuel consumed (tonnes per day to tonnes per day) ranges from 0.087 for slow speed engines to 0.057 for medium speed engines. Also directly proportional to the amount of fuel used is fuel cost, one of the most important components of total cost (although by no means the only one). Fuel cost can be estimated by multiplying the amount of bunkers used with the price of fuel. In our analysis we assume that the price of the fuel used by the ship is known and equal to p, assumed constant during the year. Even though it is assumed a constant in our analysis, p is very much market-related, and, as such, may fluctuate widely in time, as historical experience has shown (see Fig. 2 below). But this assumption causes no loss of generality, as an average price can be used. Also, as the ship will generally consume different kinds of fuels during the trip and in port, assuming a unique fuel price is obviously a simplification. But this causes no loss of generality either, as the analysis can be readily extended to account for different fuel types on board.

4. A Simple Logistical Scenario: Factors and Tradeoffs

Given that fuel costs and emissions are directly proportional to one another (both being directly proportional to fuel used), it would appear that reducing both would be a straightforward way towards a "win-win" solution. In an operational setting, one of the obvious tools for such a simultaneous reduction is speed: sail slower, and you reduce both emissions and your fuel bill. This may sound simple, but its possible ramifications are not so simple.

Assuming a given ship, and for speeds that are close to the original speed, the effect of speed change on fuel consumption is assumed cubic, that is,

$$\frac{F}{F_o} = \left(\frac{V}{V_o}\right)^3$$

where $F(F_0)$ is the daily fuel consumption at speed $V(V_0)$.

This assumption comes from basic ship hydrodynamics. It means that $F = kV^3$, where k is a known constant, which is a function of the loading condition of the ship and of other ship characteristics (e.g., engine,

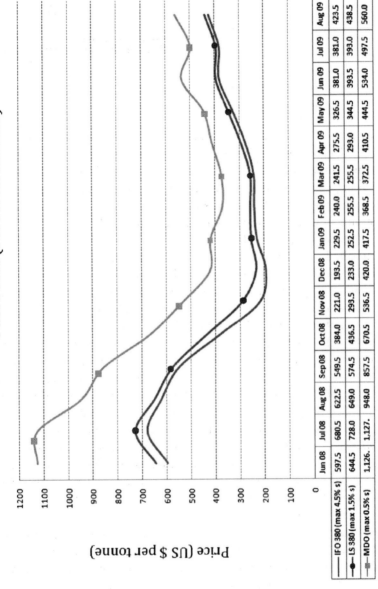

Fig. 2   Average monthly fuel oil prices (from www.bunkerworld.com).

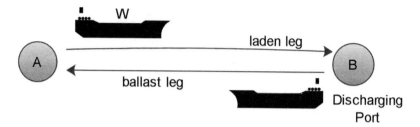

Fig. 3 Ship route.

horsepower, geometry, age, etc). Of course, an implicit assumption in this analysis is that the ship's power plant would still be able to function efficiently if speed is reduced. Speed reduction usually requires reconfiguring the engine so that its operation is optimized at the reduced load.

Also note that the cubic law is only an approximation, and one that is usually valid for small changes in speed. If the speed changes drastically, for instance from 20 to 10 or even 5 knots, one would expect a different relationship between V and F.

Our simplest logistical scenario to investigate tradeoffs between ship CO_2 emissions and other attributes of the ship operation assumes a fleet of N identical ships (N: integer), each of capacity (payload) W. Each ship loads from a port A (time in port T_A, days), travels to port B with known speed V_1, discharges at B (time in port T_B, days) and goes back to port A in ballast, with speed V_2. Assume speeds are expressed in km per day. The distance between A and B is known and equal to L (km). Assume these ships are chartered on a term charter and the charterer, who is the effective owner of this fleet for the duration of the charter, incurs a known operational cost of O_C per ship per year. This cost depends on market conditions at the time the charter is signed and includes the charter to the ship owner(s) and all other *non-fuel related* expenses that the charterer must pay, such as canal tolls, port dues, cargo handling expenses, and so on. Not included in O_C are fuel expenses, which are also paid by the charterer, and which depend on the actual fuel consumed by the fleet of ships. The latter depends on how the fleet is used.

Obviously, the above rudimentary scenario (a ship going fully laden one way and on ballast on the return leg) is not the only one that one may encounter in world shipping markets. This scenario is encountered mainly in the charter market and specifically in the tanker trades. Bulk carriers may also be employed likewise; however they are more likely to also trade in triangular routes, depending on the cargoes that are available.

Containerships and other ships in the liner market definitely do not use such employment pattern, being engaged in trades that visit many ports. Even though these operational scenarios are different from the one examined above, extending our approach to these other scenarios is straightforward, and the main thrust of our analysis is valid for these scenarios as well.

Assume that each ship's operational days per year are D $(0 < D < 365)$, a known input, and that the total daily fuel consumptions (including both main engine and auxiliaries) are known and are as follows for each ship:

In port: f (tonnes per day)
At sea: F_1, F_2 (tonnes per day) for laden and ballast legs (respectively).

As stated earlier, the effect of speed change on fuel consumption is assumed cubic for the same ship, that is, $F_{new}/F = (V_{new}/V)^3$, or, $F_1 = k_1 V_1^3$, $F_2 = k_2 V_2^3$, where k_1 and k_2 are known constants. Also as mentioned in the previous Section, one tonne of fuel burned in the ship's engine room will produce F_{CO_2} tonnes of CO_2, where F_{CO_2} is the emissions factor.

In addition to the standard costs borne by the charterer, our analysis will also take into account *cargo inventory costs*. The reason is that any conceivable speed reduction to save fuel costs and/or reduce emissions will have as a consequence an increase in inventory costs due to late delivery of cargo and must be taken into account if the analysis is to be complete from a logistical standpoint. These cargo inventory costs are assumed equal to I_C per tonne and per day of delay, where I_C is a known constant. In computing these costs, we assume that cargo arrives in port 'just-in-time', that is, just when each ship arrives. In that sense, inventory costs accrue only when loading, transiting (laden) and discharging. We shall call these inventory costs 'in-transit inventory costs'. Generalizing to the case where inventory costs due to port storage are also considered is straightforward.

If the market price of the cargo at the destination (CIF price) is P ($/tonne), then one day of delay in the delivery of one tonne of this cargo will inflict a loss of $PR/365$ to the cargo owner, where R is the cost of capital of the cargo owner (expressed as an annual interest rate). This loss will be in terms of lost income due to the delayed sale of the cargo. Therefore, it is straightforward to see that $I_C = PR/365$.

Based on the above, and on a per ship basis, and after some straightforward algebraic manipulations, we can compute the following:

Round trip duration:

$$d = L/V_1 + L/V_2 + T_{AB},$$

where

$$T_{AB} = T_A + T_B \quad \text{(total port time per round trip)}$$

Number of round trips in a year: $n = D/d$

Therefore $n = D/[L/V_1 + L/V_2 + T_{AB}]$ (note that n may not necessarily be an integer)

Total roundtrip fuel consumption: $T_{FC} = T_{AB}f + L(k_1V_1^2 + k_2V_2^2)$

[*As a parenthesis, it can be seen here that although the per day fuel consumption is a cubic function of speed, the roundtrip fuel consumption is only a quadratic function of speed, as the slower the ship goes, the more days it stays at sea.*]

Total costs in a year:

$$pnT_{FC} + nI_CW\left(T_{AB} + \frac{L}{V}\right) + O_C$$

$$= np[T_{AB}f + L(k_1V_1^2 + k_2V_2^2)] + nI_CW\left(T_{AB} + \frac{L}{V_1}\right) + O_C$$

$$= D\frac{p[T_{AB}f + L(k_1V_1^2 + k_2V_2^2)] + I_CW\left(T_{AB} + \frac{L}{V_1}\right)}{\frac{L}{V_1} + \frac{L}{V_2} + T_{AB.}} + O_C$$

Fuel consumed per tonne-km: T_{FC}/WL

For a fleet of N ships, total fleet costs in a year:

$$pnNT_{FC} + nNkW\left(T_{AB} + \frac{L}{V}\right) + NO_C$$

$$= nNp[T_{AB}f + L(k_1V_1^2 + k_2V_2^2)] + nNI_CW\left(T_{AB} + \frac{L}{V_1}\right) + NO_C$$

$$= DN\frac{p[T_{AB}f + L(k_1V_1^2 + k_2V_2^2)] + I_{CW}\left(T_{AB} + \frac{L}{V_1}\right)}{\frac{L}{V_1} + \frac{L}{V_2} + T_{AB.}} + NO_C$$

With this basic scenario complete, we are now ready to investigate the impact of speed reduction.

To investigate what happens if we reduce speed, we assume that we reduce the speed of all ships in the fleet by a common amount.[mm] Let this common reduction (initial speed — final speed) be equal to $\Delta V \geq 0$.[nn]

[mm]Reducing speeds by different amounts is a straightforward generalization.
[nn]We implicitly assume that we shall not consider a speed increase, or $\Delta V < 0$, even though this may be warranted cost-wise. A speed increase will always increase fuel

To reduce speed and maintain annual throughput constant, we have to add more ships. If these additional ΔN ships are identical in design to the original N ones, ΔN can be determined by equating nNW (the quantity of cargo moved in a year with N ships) with the equivalent expression for $N + \Delta N$ ships. ΔN may not necessarily be an integer, although for illustration purposes one may want to round it to the next highest integer.

It is easy to check that we can compute ΔN from the following equation:

$$\Delta N = N \left(\frac{\frac{L}{V_1 - \Delta V} + \frac{L}{V_2 - \Delta V} + T_{AB}}{\frac{L}{V_1} + \frac{L}{V_2} + T_{AB}} - 1 \right)$$

Before we proceed, we implicitly assume that these ΔN ships are readily available and can be immediately incorporated into the original fleet at a cost equal to O_C per ship per year, the same as that paid to charter the original N ships. However, this may not be the case if there is a lack of supply of available ships, which may have as a result a lower total throughput and/or an increase of charter rates to levels above O_C. Also, and as we investigate an operational setting, we do not take into account long-term effects such as emissions produced by shipyards that would build these extra ships, emissions produced by the ships carrying the additional raw materials to be used to build these ships, and other similar life-cycle quantities.

After some straightforward algebraic manipulations, the difference in total fleet costs (costs after, minus costs before) is equal to

$$\Delta(total\ fleet\ costs) = NL\Delta V$$
$$\times \frac{-pD(2k_1 V_1 + 2k_2 V_2 - (K_1 + k_2)\Delta V) + \frac{I_C W D}{V_1(V_1 - \Delta V)} + O_C \left(\frac{1}{V_1(V_1 - \Delta V)} + \frac{1}{V_2(V_2 - \Delta V)} \right)}{\frac{L}{V_1} + \frac{L}{V_2} + T_{AB}}$$

$$(1)$$

Or, in simplified form, if $V_1 = V_2 = V$ (this may not mean that $k_1 = k_2$):

$$\Delta(total\ fleet\ costs)$$
$$= NL\Delta V \frac{-pD(2V - \Delta V)(k_1 + k_2) + \frac{I_C W D + 2 O_C}{V(V - \Delta V)}}{2\frac{L}{V} + T_{AB}} \qquad (2)$$

---

consumption and emissions, but may actually entail lower other costs, such as inventory or other, leading in turn to lower total costs.

The difference in fuel costs alone (costs after minus costs before) is equal to

$$\Delta(\text{total fuel costs}) = -NL\Delta V \frac{pD(2k_1V_1 + 2k_2V_2 - (k_1 + k_2)\Delta V)}{\frac{L}{V_1} + \frac{L}{V_2} + T_{AB}} \quad (3)$$

Or, in simplified form,

$$\Delta(\text{total fuel costs}) = -NL\Delta V \frac{pD(2V - \Delta V)(k_1 + k_2)}{2\frac{L}{V} + T_{AB}} \quad (4)$$

An interesting observation is that fuel cost differentials (and, by extension, total fleet cost differentials) are independent of port fuel consumption f. Even though this may seem counter-intuitive, it can be explained by noting that the new fleet string, even though more numerous than the previous one, will make an equal number of port calls in a year, therefore fuel burned in port will be the same.

It is also interesting to note that for $\Delta V \geq 0$ and for all practical purposes the differential in fuel costs is always negative or zero, as the term within the square brackets of [3], or the difference $2V - \Delta V$ in [4], is positive for all realistic values of the speeds and of the speed reduction. This means that speed reduction cannot result in a higher fuel bill, even though more ships will be necessary.

The same is true as regards emissions, as these are directly proportional to the amount of fuel consumed:

$$\Delta(\text{total CO}_2 \text{ emissions})$$
$$= -F_{CO_2}NL\Delta VD \frac{(2k_1V_1 + 2k_2V_2 - (k_1 + k_2)\Delta V)}{\frac{L}{V_1} + \frac{L}{V_2} + T_{AB}} \quad (5)$$

Or, in simplified form,

$$\Delta(\text{total CO}_2 \text{ emissions}) = -F_{CO_2}NL\Delta VD \frac{(2V - \Delta V)(k_1 + k_2)}{2\frac{L}{V} + T_{AB}} \quad (6)$$

Total emissions would thus be always reduced by slowing down, even though more ships would be used. The higher the speed, and the higher the speed reduction, the higher this reduction would be.

As a parenthesis we note that mathematically expression [6] achieves its lowest value (that is, emissions reduction is maximized) if $\Delta V = V$. This option is of course only of theoretical value, for if this is the case the fleet would come to a complete standstill and the other cost components (as well as ΔN) would go to infinity.

In the general case, whether Δ(total fleet cost) in expressions [1] or [2] is positive or negative, or reaches a minimum value other than zero, would depend on the values of all parameters involved, for one can see that in-transit inventory costs and ship other operational costs count positively in the cost equation. Both these costs would increase by reducing speed, and this increase might offset, or even reverse, the corresponding decrease in fuel costs. High values of either I_C or O_C (or both) would increase the chances of this happening, and high values of p would do the opposite, as will be seen in the examples that follow.

A closer look at expression [2][oo] provides some interesting insights. Expression [2] can be written in the following form:

$$\Delta(\text{total fleet cost}) = \Delta V \left(-A(2V - \Delta V) + \frac{B}{V - \Delta V} \right) \equiv G(\Delta V)$$

where A and B are positive constants given by:

$$A = NL_p D \frac{(k_1 + k_2)}{2\frac{L}{V} + T_{AB}} \quad B = NL \frac{I_C W D + 2O_C}{V \left(2\frac{L}{V} + T_{AB} \right)}$$

As we have assumed that $\Delta V \geq 0$, function $G(\Delta V)$ obtains the value of 0 for $\Delta V = 0$ and goes to infinity when ΔV approaches V. Its behavior for intermediate values of ΔV depends on the values of all parameters involved. In fact, we distinguish two cases:

Case 1: The derivative of G(ΔV) at $\Delta V = 0$ is ≥ 0 (see Fig. 4a below). This is mathematically expressed as $V \leq \sqrt{\frac{B}{2A}}$, or as

$$V \leq V_0 \text{ with } V_0 \equiv \sqrt[3]{\frac{I_C W + 2\frac{O_C}{D}}{2p(k_1 + k_2)}} \tag{7}$$

Speed V_0 depends on the parameters shown above and can be considered as a cost-benefit 'speed threshold'. If the original speed of the ship V is at or below that threshold, then any attempt to reduce it to save fuel (and emissions) would entail a net total cost increase, as $G(\Delta V)$ will be monotonically increasing with ΔV.[pp] It can be seen that this situation is more likely to occur if I_C and/or O_C are high and/or p is low.

[oo]The analysis for expression [1] is similar, but more tedious.
[pp]Again, in this case it may be argued that it is best to increase speed, and reduce the number of ships, or that $\Delta V < 0$. But this is a case that was excluded from the beginning.

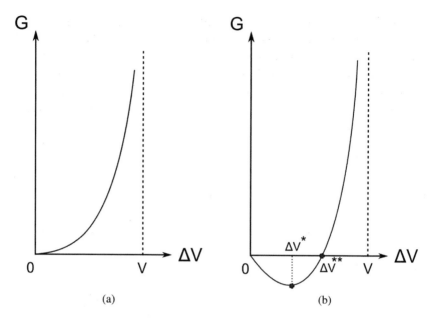

Fig. 4   Possible forms of $G(\Delta V)$.

Case 2: The derivative of $G(\Delta V)$ at $\Delta V = 0$ is < 0 (see Fig. 4b).

This is mathematically expressed as $V > \sqrt{\frac{B}{2A}}$, or as $V > V_0$ with V_0 defined as in [7] above.

If the original speed of the ship V is above the V_0 threshold, then the option to reduce speed to save fuel (and emissions) could also reduce total costs. This situation is more likely to occur if I_C and/or O_C are low and/or p is high.

In this case, $G(\Delta V)$ achieves a minimum (negative) value for some 'optimal' value of $\Delta V = \Delta V^*$, between 0 and V. In fact, $G(\Delta V) \leq 0$ for $0 \leq \Delta V \leq \Delta V^{**}$, and $G(\Delta V) > 0$ for $\Delta V > \Delta V^{**}$, where ΔV^{**} is the other (nonzero) root of $G(\Delta V) = 0$. We note that $\Delta V^{**} > \Delta V^*$. Both ΔV^* and ΔV^{**} depend on the values of all other parameters.

If this is the case, speed reduction would indeed be beneficial, and choosing $\Delta V = \Delta V^*$ would achieve maximum total benefits.

We now present several simple examples to illustrate our approach.

Example 1 — Aframax Tanker Fleet

The first example considers a fleet of $N = 10$ Aframax double hull tankers, each with a DWT of 106,000 tonnes, and payload $W = 90,000$ tonnes, serving the route from Ras Tanura to Singapore, a distance of $L = 3,702$ nm

(6,871 km). Other input parameters are as follows:

$V_1 = V_2 = 15\,\text{knots} = 668.16\,\text{km/day}.$

$T_A = T_B = 4\,\text{days}$

$F_1 = F_2 = 65\,\text{tonnes/day}$ (meaning that $k_1 = k_2 = 2.1791 \cdot 10^{-7}$

$D = 350\,\text{days}$

$f = 50\,\text{tonnes/day}$

$p = \$218/\text{tonne}$ (December 2008)

$p = \$600/\text{tonne}$ (July 2008)

In other words, we examine two variants, one with a low fuel price and one with a high one (all else being equal).

Then we consider reducing speed by one knot, to 14 knots, or 623.62 km/day. It is straightforward to show that we will need 0.60 more ships to be able to cover the same annual throughput. Rounding off to one more ship, we will have (Table 3):

We can see that fuel costs are reduced in both variants, the cost differential being $512,541 in the low fuel price variant and $1,410,663 in the high fuel price variant, both on a yearly basis. CO_2 averted would amount to 54,400 tonnes, even though one more ship is employed.

Still, this does not necessarily mean that total fleet costs will be reduced, as these would also depend on inventory and other operational costs.

Neglecting inventory costs for this example (these will be examined in example no. 3), we consider what the other operational costs might be in each of these variants.

In a market as seriously depressed as in late 2008, ship owners have been said to be willing to charter their ships for a rate of zero, with the charterer paying only for fuel. In this case, variant 1 would continue to be

Table 3   Aframax tanker comparison.

Quantity	10 ships going 15 knots	11 ships going 14 knots
Total fuel consumed for fleet, (tonnes per year)	218,952	201,778
CO_2 for fleet (tonnes per year)	694,077	639,637
Bunker cost for fleet ($/year)		
Fuel price $p = 218$ $/tonne	$7,143,419	$6,630,878
$p = 600$ $/tonne	$19,660,787	$18,250,124

profitable, although the net savings, if expressed per day, would be very meager ($1,404/day).

For the high-market variant however, the $3,865/day savings of fuel costs are well below what an Aframax could command when the market was high. Rates as high as $60,000/day have been observed for this type of ship (or perhaps even higher), meaning that speed reduction during these periods would be non-sensical from a cost-benefit viewpoint.

Example 2 — Panamax Containership Fleet

Our second illustrative example investigates the effect of speed reduction in containerships. As said earlier, containerships are the top CO_2 emissions producer in the world fleet (2007 Lloyds-Fairplay database).

Assuming a hypothetical string of $N = 100$ (identical) Panamax containerships, each with a payload of $W = 50,000$ tonnes, if the base speed is $V = 21$ knots (both ways) and fuel consumption at that speed is 115 tonnes/day, then for a fuel price of $p = $600/tonne (corresponding to a period of high fuel prices, before the slump of 2008), the daily fuel bill would be $69,000 per ship. Running the same type of ship at a reduced speed $V - \Delta V = 20$ knots (one knot down), the fuel consumption would drop to 99.34 tonnes/day (cube law vs. 21 knots) and the daily fuel bill would drop to $59,605 per ship, some $10,000/day lower.

Assume these 100 ships go back and forth a distance of 2,100 miles (each way) and are 100% full in one direction and completely empty in the other. This is not necessarily a realistic operational scenario, as containerships visit many ports and as capacity utilizations are typically lower both ways, depending on the trade route. The scenario of trade routes from the Far East to Europe or from the Far East to North America, which are almost full in one direction and close to empty in the other, is probably close to the assumed scenario. However, a generalization of this analysis to many ports and different capacity utilizations in each leg of the trip should be straightforward. For simplicity, assume $D = 365$ operating days per year

Table 4   Panamax containership comparison.

Quantity	100 ships going 21 knots (case A)	105 ships going 20 knots (case B)
Total fuel consumed for fleet, (tonnes per year)	4,197,500	3,807,256
CO_2 for fleet (tonnes per year)	13,306,075	12,069,002
Bunker cost for fleet ($/year)	2,518,500,000	2,284,353,741

and zero loading and unloading times. For non-zero port times, the analysis will be more involved but will lead to similar results.

At a speed of 20 knots, we will need 105 ships to reach the same throughput per year. Then we will have:

The net reduction of CO_2 emissions (per year) is 1,237,073 tonnes, and the fuel cost reduction (per year) is $234,146,259 for 5 more ships, that is, $46,829,252 per additional ship. Dividing by 365, this difference is $128,299 per day.

This means that if the sum of additional cargo inventory costs plus other additional operational costs of these ships (including the time charter) is less than $128,299 a day, then case B is overall cheaper. One would initially think that such a threshold would be enough. But it turns out that this is not necessarily the case if in-transit inventory costs are factored in.

Before we do so, we display Table 5, that illustrates the unit value of the top 20 containerized imports at the Los Angeles and Long Beach Ports in 2004 (see CBO(2006)).

Table 5   Unit value of containerized imports (1,000 $ per short ton).

Unit value of the top 20 containerized imports at los angeles and long beach ports, 2004

HS#	Category of import	Value (billions of dollars)	Weight (thousands of short tons)	Unit Value (thousands of dollars per ton)
84	Machinery, boilers, reactors, parts	38.0	698.6	54.3
85	Electric machinery, sound and television equipment, parts	31.7	677.0	46.8
87	Vehicles and parts, except railway or tramway	12.1	337.4	35.8
62	Apparel articles and accessories, not knit or crochet	9.9	132.4	74.6
95	Toys games, and sports equipment and parts	9.4	377.1	25.0
94	Furniture, bedding, lamps, etc.	9.3	739.8	12.6
61	Apparel articles and accessories, knit or crochet	9.0	132.1	68.4
64	Footwear	7.8	181.4	43.0
39	Plastics and articles thereof	5.2	409.0	12.8
73	Articles of iron or steel	4.4	467.0	9.4
42	Leather articles, saddlery, handbags	3.8	117.2	32.1
90	Optic, photographic, and medical instruments	3.6	41.8	86.2

Note that one short ton is equal to 0.9072 tonnes.

To compute in-transit inventory costs for the above example, we hypothetically assume that cargo carried by these vessels consists of high value, industrial products, similar to those in Table 5, and that its average value at the destination (CIF price) is $20,000/tonne. We also assume the cost of capital being 8%. This means that one day of delay of one tonne of cargo would entail an inventory cost of $I_C = PR/365 = 20,000*0.08/365 = 4.38. This may not seem like a significant figure, but it is.

Computing the in-transit inventory costs for this case gives a total annual difference of $200,000,000 ($4,200,000,000–$4,000,000,000) in favor of case A, which moves cargo faster. This figure is significant, of the same order of magnitude as the fuel cost differential.

Assuming also a time charter rate of $25,000 per day (typical charter rate for a Panamax containership in 2007), the total other operational costs of the reduced speed scenario are $958,125,000 per year for 105 ships, versus $912,500,000 for 100 ships going full speed. Tallying up we find a net differential of $11,478,741 per year in favor of case A, meaning that in-transit inventory and other operational costs offset the positive difference in fuel costs.

Of course, other scenarios may yield different results, and the reduced speed scenario may still prevail in terms of overall cost, under different circumstances. For instance, if the average value of the cargo is $10,000/tonne, and everything else is the same, then the difference in annual inventory costs drops to $100,000,000, rendering the reduced speed scenario a profitable proposition (with a total cost reduction of $88,521,259 per year). Actually, speed reduction remains profitable if the value of the cargo is no more than about $18,800/tonne (which can be considered as a break-even CIF price).

All of the above confirm that the drive to reduce emissions may or may not be a win-win proposition, with the final outcome depending on the specific parameters of the particular scenario (see Psaraftis and Kontovas (2000b) for some additional insights).

We end this section by noting that there are cases where adding more ships may not be necessary. These are cases in which the ship's schedule by design includes an amount of idle time in port. Such cases are typical for RoPax scheduled operations, where there is idle time built into the ship's schedule for various operational reasons. In these cases, any delay due to speed reduction is absorbed by the available idle time and no additional ships are necessary. For a discussion of this scenario, see Psaraftis et al. (2009c).

5. The Cost to Avert One Tonne of CO_2

What would it take to avert one tonne of CO_2 by speed reduction? Or, put in a different way, as much as the question "what price safety?" is common, let us now ask "what price emissions reduction?" We address this question by noting that in expressions [5] and [6], Δ (total CO_2 emissions) equals minus total CO_2 averted by implementing a speed reduction scheme. We define as the cost to avert one tonne of CO_2 (CATC) the ratio of the total net cost of the fleet due to CO_2 speed reduction divided by the amount of CO_2 averted by speed reduction. Then we will have:

$$CATC = NL\Delta V \frac{-pD(2V - \Delta V)(k_1 + k_2) + \frac{I_C W D}{V(V-\Delta V)} + \frac{2O_C}{V(V-\Delta V)}}{F_{CO_2} NDL\Delta V(2V - \Delta V)(k_1 + k_2)}$$

After some algebraic manipulations, this can be rewritten as

$$CATC = \frac{I_C W D + \frac{2O_C}{D}}{F_{CO_2} V(V - \Delta V)(2V - \Delta V)(k_1 + k_2)} - \frac{p}{F_{CO_2}} \tag{8}$$

It can be seen that CATC is a positive linear function of both I_C and O_C and a negative linear function of the price of fuel p. It can also be seen that the denominator in the bracket is a cubic function of speed, reflecting the functional relationship between speed and the quantity of CO_2 that is produced.

In addition, the last term in [8], $-p/F_{CO_2}$, where p is the price of one tonne of fuel and F_{CO_2} is the CO_2 emissions factor, can be recognized as the cost of the amount of fuel saved (not spent) that would produce one tonne of CO_2. This is an opportunity cost that we will have to subtract from the total cost incurred, as it corresponds to the amount of fuel that would be saved if one tonne of CO_2 is averted.

The CATC criterion can be used whenever alternative options to reduce emissions are contemplated. In that sense, the alternative that achieves the lowest CATC is to be preferred.

The case in which CATC is negative corresponds to the case in which reducing speed is cost-beneficial, that is, to the case the function $G(\Delta V)$ of the previous section takes on a negative value.

For the containership example of the previous section, the CATC values for the various scenarios examined are as follows (Table 6):

This table confirms that CATC can vary widely. It is also interesting to note that the difference in CATC between the 1st and 2nd scenario is the same as that between the 3rd and 4th scenario ($80.84/tonne in both

Table 6   Values of CATC as per containership
scenarios outlined earlier.

Scenario	CATC ($/tonne of CO_2 averted)
p = $600/tonne P = $20,000/tonne OC = $25,000/day	9.28
p = $600/tonne P = $10,000/tonne OC = $25,000/day	−71.56
p = $250/tonne P = $20,000/tonne OC = $15,000/day	104.94
p = $250/tonne P = $10,000/tonne OC = $15,000/day	24.10

cases). This is not a coincidence, and can be explained by the structure of expression [8].

In these examples, the influence of in-transit inventory costs in the value of CATC can also be seen clearly. This means that perhaps one of the biggest obstacles that needs to be overcome if emissions are to be reduced, is the unwillingness of the cargo owners to incur inventory costs for their cargoes. Optimized routing, logistics, and other operational measures that would reduce this inventory costs would be important.

As regards what threshold conceivably exists for CATC, that is, under what (positive) value of CATC a speed reduction scheme would still be considered desirable, this issue is currently open and it is not an easy one to address. As much as it is obvious that both the shipping community and society at large wish to reduce CO_2 emissions from shipping, it is far from clear how much they are willing to pay to do so. This is not a surprise, given the fact that there is wide disparity of views on what should be done to curb GHG emissions, and the fact that decisions on the CO_2 front are still pending.

In a conceivable CO_2 Emissions Trading Scheme (ETS) for shipping, a monetary value would be put on a per tonne basis, for instance, $30/tonne of CO_2 averted, and emissions reduction measures would be evaluated against such a threshold. Such market values for CO_2 currently exist for other industries, but not for shipping, for which it is unclear how, or when such a scheme would be implemented.

ECX EUA Futures Contracts

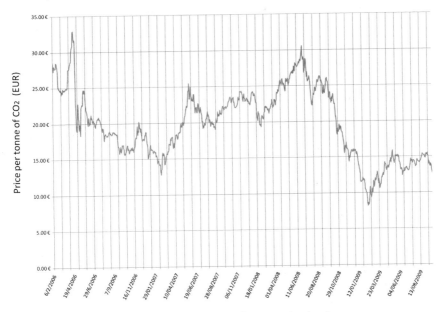

Fig. 5   ECX EU allowances future contract prices.
Source: European Climate Excange.

Figure 5 shows the historic 2009 settlement prices of EU allowances issued under the EU Emissions Trading Scheme and traded at the European Climate Exchange (ECX). One EUA equals one tonne of CO_2 (right-to-emit).

The concept of CATC, as defined above, can be generalized to measures other than speed reduction, and can be a useful concept for the evaluation of policy or other alternatives.

6. The Port Time Factor

This section focuses on the case where total trip time is kept constant, even though speed is reduced (see Psaraftis *et al.* (2009c) for more details). Given the fact that time at sea increases with slow steaming we must investigate possible ways to decrease time in port. This is not an easy task. The most feasible way to reduce time in port is through operational decisions regarding land-side operations (berth allocation, quay cranes scheduling and vessel stowage). Optimizing terminal operations has received increasing interest over the last years. Vis and de Koster (2003) review the relevant literature and illustrate the main logistics processes in a container terminal whereas Steenken *et al.* (2004) provide an overview of optimization methods terminal operations. The problem of allocating ships to berths (discrete case) or to quays (continuous case) is dealt among others in Cordeau *et al.* (2005) and Wang and Lim (2007). The Quay Crane Scheduling Problem (QCSP) which refers to the allocation of cranes and to the scheduling of stevedoring operations can be solved with the use of dynamic programming as proposed in Lim *et al.* (2004) or be addressed with a greedy randomized adaptive search procedure like the one analyzed in Kim and Park (2004). Lee *et al.* (2006) address a yard storage allocation problem to reduce traffic congestion and Lee and Hsu (2007) present model for container re-marshalling. For a circumstantial review of the operations research literature of problems related to container terminal management the reader could refer among others to Vis and de Koster (2003) and Steenken *et al.* (2004).

We now present a simple scenario to investigate the impact of speed reduction on ship CO_2 emissions and fuel costs in the case that the total trip time is kept constant.

Assume a ship that loads from a port A, travels to port B (a total distance of L nm from A) carrying a payload W with a known speed of V_0 (in knots), where she discharges the cargo and stays at port before departing again.

The daily fuel consumptions and times that the ship spends at sea and in port are known and are as follows:

At sea:

Fuel consumption F_0 (tonnes per day)
Total time at sea $T_0 = \frac{L}{24 \cdot V_0}$ (days)

In port:

Fuel consumption f (tonnes per day)
Total time in port t_0 (days)

Thus, the total fuel consumption for this trip is $FC_0 = F_0 \cdot T_0 + f \cdot T_0$.

Now suppose that the ship operator wants to investigate the scenario of speed reduction. The new speed V will be a fraction of the original speed ($V = aV_0$ where $0 < a < 1$) and, hence, there will be an increase of the time at sea, $T = \frac{L}{24V} = \frac{T_0}{a}$

The effect of speed change on fuel consumption is assumed cubic for the same ship (and for speeds that are close to the original speed) as discussed in Section 4. The fuel consumption in port per day will remain the same, but we assume that the new time in port (t) will be reduced in order to keep at least the same total trip time with that before the speed reduction.

For this trip we can compute the difference in fuel consumption as follows:

Δ(Fuel consumption)

$$= \Delta(\text{consumption at sea}) + \Delta(\text{consumption at port})$$

$$= F \cdot T - F_0 \cdot T_0 + f \cdot t - f \cdot t_0$$

$$= F \cdot \frac{L}{24 \cdot V} - F_0 \cdot \frac{L}{24 \cdot V_0} + f \cdot (t - t_0)$$

$$\overset{V = aV_0}{=} F \cdot \frac{L}{24 \cdot aV_0} - F_0 \cdot \frac{L}{24 \cdot V_0} + f \cdot (t - t_0)$$

$$= \frac{L}{24 \cdot V_0} \left(F \cdot \frac{1}{a} - F_0 \right) + f \cdot (t - t_0)$$

$$\overset{\frac{F}{F_0} = \left(\frac{V}{V_0}\right)^3}{=} \frac{L}{24 \cdot aV_0} \left(\left(\frac{V}{V_0}\right)^3 F \cdot \frac{1}{a} - F_0 \right) + f \cdot (t - t_0)$$

$$= \frac{L}{24 \cdot V_0} \left(\left(\frac{aV_0}{V_0}\right)^3 F \cdot \frac{1}{a} - F_0 \right) + f \cdot (t - t_0)$$

$$= \frac{L}{24 \cdot V_0} \left(a^3 F_0 \cdot \frac{1}{a} - F_0 \right) + f \cdot (t - t_0)$$

Thus, the total fuel consumption for slow steaming is:

$$\Delta(\text{Fuel consumption}) = \frac{L}{24 \cdot V_0} F_0 (a^2 - 1) + f \cdot (t - t_0) \qquad (9)$$

As one may notice, the first addend is negative since, by definition, parameter 'a' lies between 0 and 1 and L, F_0 and V_0 are always positive. It is obvious that if time in port remains the same ($t - t_0$ equal to 0) there will be a need to add a number of additional vessels (possibly fractional) in order to maintain the same throughput per year. The model in this Section examines assumes that $t < t_0$, and in fact that 't' is such that the total trip time, including time in port, remains the same ($T + t = T_0 + t_0$).

Furthermore, as discussed in Section 3, to find the equivalent CO_2 emissions reduction, one has to multiply the reduction in bunker consumption by the appropriate emissions factor (F_{CO_2}) from Table 2.

$$\Delta(CO_2 \text{ emissions}) = F_{CO_2} \cdot \left[\frac{L}{24 \cdot V_0} F_0(a^2 - 1) + f \cdot (t - t_0) \right] \qquad (10)$$

Again, the fuel cost reduction can be estimated by assuming a constant fuel price as in the previous examples.

$$\Delta(\text{fuel costs}) = p \cdot \left[\frac{L}{24 \cdot V_0} F_0(a^2 - 1) + f \cdot (t - t_0) \right] \qquad (11)$$

We now move forward to a realistic example using the following figures that are based on operational data provided by Det Norske Veritas (DNV), see Psaraftis *et al.* (2009c).

A Panamax container-vessel begins its trip at Port A, and then consequently visits ports B and C before going back. The time that she spends at sea and in port and the relating fuel consumptions are as follows:

Depart port	Arrive port	Distance (miles)	Avg speed (kn)	Total TEU	Sailing time (hrs)	F_0 (tn/ day)	T_0 (days)	f (tn/ day)	t_0 (days)
A	B	115	20.18	1892	5.70	91.79	0.24	16.58	1.79
B	C	6068	23.41	2593	259.20	136.81	10.80	3.26	5.45
C	A	6323	22.85	3294	276.70	139.22	11.53	12.15	3.55

Using Eqs. 9,10,11 and we can calculate the reductions in fuel cost and emissions for each leg. Note that the emissions factor used in this example is 3.13.

For reasons of simplicity we omit the detailed calculations and we present the resulting total reductions for this round trip in Fig. 6.

One can observe some significant savings in fuel consumption, CO_2 emissions (in fact, all emissions) and fuel cost. However, "there is no free lunch" necessarily. Compensating for a reduced speed will entail either additional ships to maintain the same throughput, or the ability to reduce port time. If the former can be achieved, overall emissions are shown to be reduced, but the overall cost (including cargo in-transit inventory cost) may or may not go down (as per previous section). Emissions can be reduced even further if port time can be reduced so that there is no need for additional vessels. But this may be a more difficult proposition.

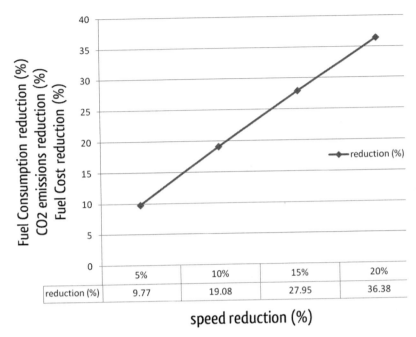

Fig. 6   Reductions in fuel consumption, CO2 emissions and Fuel Costs.

For instance, in the example illustrated above, when speed is reduced by 5%, time in port has to be reduced by 11% to maintain a constant total trip time. If this sounds feasible, it is non-trivial nonetheless. For a speed reduction of 15% the total time in port has to be reduced from 10.8 days down to 6.81, which is almost a 37% reduction. This is a much more difficult proposition, possibly entailing drastic port re-engineering and/or infrastructure improvements.

7. Speed Reduction at SECAs

All considerations of the previous sections of this paper can be also applied to emissions other than CO_2. For instance, one can compute emissions from other pollutants and also define CATN (the cost to avert one tonne of NOx), CATS (the cost to avert one tonne of SO_2),[qq] and so on.

[qq]Not to be confused with CATS (cost to avert one tonne of spilled oil) — a criterion under discussion in Formal Safety Assessment (in the context of environmental risk evaluation criteria).

The only difference with the previous analysis is that one would have to substitute for the CO_2 emissions factor the appropriate emissions factor of the pollutant under consideration. For instance if one considers SO_2, one has to multiply total bunker consumption (in tonnes per day) by the percentage of sulphur present in the fuel (for instance, 4%, 1.5%, 0.5%, or other) and subsequently by a factor of 0.02 to compute SO_2 emissions (in tonnes per day). Even though the amounts of SO_2 produced by ships are substantially lower than CO_2, for SO_2 emissions other considerations are equally important. SO_2 is not a greenhouse gas but as it causes acid rain (among other effects), its reduction is a matter of high priority. To that effect, SO_2 (and generally SO_x) reduction is also high on the IMO agenda, and in fact regulatory progress on this front is more advanced than for the CO_2 front, as exemplified in the latest MEPC 58 Annex VI developments as regards the timetable on SO_x emissions caps.

To reduce pollution by SO_x, special highly sensitive areas have been designated by the IMO as 'Sulphur Control Emissions Areas', or SECAs, where specific limits in SO_x content are set for a ship's exhaust gases. Designated SECAs to date are the Baltic Sea, the North Sea and the English Channel. The IMO does not specify how the SO_x emissions targets should be reached. Among methods contemplated, sea scrubbers are a measure that is offered on the technology front. Fuels cleaner in sulphur content is also a method that is proposed (of which more later).

Among potential operational measures, one question that is relevant is this: Can speed reduction at SECAs work, as a measure to reduce SO_x emissions? This sounds like an easy question to pose, for which however the answer may not be so easy.

First of all it should be noted that speed reduction, in and of itself, will not change the proportion of SO_2 in a ship's exhaust. But it will change the total amount of SO_2 produced, much in the same way as this happens for CO_2. In that sense, speed reduction to reduce SO_2 is worthy of note.

Let us assume a ship that goes from port X to port Y, sailing a total distance of L. At the beginning or the end of the trip, there is a SECA, of distance d ($< L$).

Assume there are two options: The first (option A) is to sail the entire trip at a constant speed of V. The second (option B) is to reduce speed to v ($< V$) within the SECA, so as to reduce SO_2 emissions, but go at a slightly higher speed of V^* ($> V$) outside the SECA, so that *total transit time is the same*.

Total transit time is kept the same so that we do not need more ships in the supply chain, and shippers do not lose money on in-transit inventory costs. If total transit time is not the same, we shall have to go through an analysis similar to that of previous sections.

Let us now pose the question, with total transit time being the same, which option burns less fuel, A or B? The one that does so would also cost less, and would also produce less total emissions, not only in SO_2, but also CO_2 and all other pollutants.

The analysis is straightforward and goes as follows:

Let the transit time in both scenarios be $T = L/V$ (in days). If within the SECA the speed is v ($< V$) for distance d, then

$$\frac{L}{V} = \frac{d}{V} + \frac{L-d}{V^*}$$

Therefore

$$V^* = \frac{L-d}{\frac{L}{V} - \frac{d}{v}} \ (> V)$$

[*the assumption here is that $L/V > d/v$, otherwise making up the time lost in the SECA would be impossible.*]

Again, we assume that fuel consumption per day obeys a cube law, that is, is equal to kV^3. Since we have to multiply by total days, the law becomes quadratic, as total fuel consumption is $T_{FC} = kV^3(L/V) = kLV^2$.

Option A: total fuel consumption $T_{FC}(V) = kLV^2$
Option B: total fuel consumption $T_{FC}(V^*, v) = k(L-d)V^{*2} + kdv^2$

Substituting, we get

$$T_{FC}(V^*, v) = \frac{k(L-d)^3}{\left(\frac{L}{V} - \frac{d}{v}\right)^2} + kdv^2$$

Define the ratio

$$R = \frac{T_{FC}(V^*, v)}{T_{FC}(V)} = \frac{(L-d)^3}{L\left(L - d\frac{V}{v}\right)^2} + \frac{d}{L}\left(\frac{v}{V}\right)^2$$

It can be shown mathematically that always $R > 1$ (assuming again that $L/V > d/v$). The proof of this is straightforward.

Let us illustrate this with an example:

Let $L = 2000$ nautical miles

$d = 200$ (SECA)

$V = 20$ knots

$v = 18$ knots within the SECA

$$V^* = \frac{1800}{\frac{2000}{20} - \frac{200}{18}} = 20.25 \text{ knots outside SECA}$$

Then

$$R = \frac{1800^3}{2000 \left(2000 - 200\frac{20}{18}\right)^2} + \frac{200}{2000} \left(\frac{20}{18}\right)^2 = 0.9226 + 0.081 = 1.0036$$

Other speed and distance combinations will produce other ratios, but all will be >1.

The conclusion from this analysis is this: Speed reduction in SECAs will reduce emissions (of all gases, including SO_x) within the SECA, but result in more total emissions and more total fuel spent if speed is increased outside the SECA to make up for lost time. The reduced emissions within the SECA will be more than offset by higher emissions outside (for all gases). The total fuel bill will also be higher.

Of course, whether or not society may mind polluting the areas outside SECAs more in order to make conditions in SECAs more friendly to the environment is a non-trivial issue that is outside the scope of this paper.

Alternatively, if a lower speed is maintained throughout the ship's journey, then obviously total fuel and total emissions will be reduced, but there will be increased costs in the form of more ships needed to carry the same cargo in a year and more in-transit inventory cost for the shippers (as per previous sections).

8. SECAs Continued: Effect on Modal Split

We now make a cursory investigation of the case in which a ship involved in short sea trades uses low- sulphur fuel at a SECA, to reduce SO_x emissions. This fuel is 4–30% more expensive than high-sulphur fuel (see Fig. 3). Hence freight rates may go up. Furthermore, according to a document submitted by INTERFERRY to MEPC 58 (doc. MEPC 58/5/11) the rise in fuel prices over the past years and the cost increase for low sulphur fuel will either be passed on to customers or will force some operators out of the market. Increases in fuel prices are cited in this document as key reason for canceling certain ferry routes including those from Newcastle (United Kingdom) to Bergen, or Kristansand (Norway). This may induce shippers to use land transport alternatives (trucking), which will go against stated policies toward shifting cargo from land to sea and increase CO_2 emissions through the logistics chain. The European Community Shipowners' Association (ECSA) has already warned that new sulphur limits agreed at the IMO could push more freight onto the roads in Europe (Lloyds List, 2008a).

In this paper we shall only examine a hypothetical and rudimentary example of this scenario, which goes as follows. A modern Handymax bulk carrier moves a cargo of $W = 45,000$ tonnes from Bergen to Oslo, Norway, a distance of $L = 371$ nm (689 km). The ship sails with a speed of 14 knots and consumes 30 tonnes of HFO per day.

The ship completes the trip in 1.1 days, after consuming a total of 33.13 tonnes of fuel. Thus total CO_2 emissions are 105.01 tonnes, which corresponds to 3.39 grams per tonne-km. Total SO_2 emissions amount to 2.99 tonnes for high-sulphur (4.5%) heavy fuel oil but only to 0.33 tonnes for low-sulphur (0.5%) marine diesel oil, which is the maximum allowable sulphur content effective 1/1/2020. This means that the potential savings in SO_2 emissions by switching to cleaner fuel are 2.66 tonnes (CO_2 would be the same).

We obviously have no way of knowing what the fuel prices will be in 2020, and in particular what the availability of low sulphur fuel might be and how it could impact these fuel prices at that time. However, let us assume that prices (in 2008 US dollars) are as they were in July 2008, when they were high (see Fig. 3). Then the total bunker cost for this trip is $22,545 when using high-sulphur HFO and $37,354 for low-sulphur MDO.

Suppose now that a (yet unspecified) portion, or even the whole amount of this cargo is transported from Bergen to Oslo by road, using a modern and environment-friendly truck with a trailer (long-haul traffic) whose engine emits 2.6 kgr of CO_2 per liter of low-sulphur diesel fuel (10 ppm of sulphur).

In this case one truck moves a cargo of 40 tonnes with a speed of 60 km/h and a fuel consumption of 43 liters per 100 km when loaded. Each one-way trip from Bergen to Oslo, a distance of 490 km by road, takes 8.2 hrs or 0.34 days. Total fuel consumption is 0.2107 tonnes per one-way trip, which corresponds to 0.548 tonnes of CO_2 per one-way trip, or 27.95 grams of CO_2 per tonne-km.

We first notice that the comparison is not on a completely equal basis, as the sea trip distance is some 40% longer than the road one. Even so, let us calculate the total CO_2 produced by the road option.

To move the whole cargo of 45,000 tonnes of one shipload one way by road, it would take 1,125 truck trips, bringing the total CO_2 produced by this option to 616.3 tonnes, almost 6 times as much as that produced by the ship, and more than 230 times the amount of SO_2 potentially saved by the cleaner ship fuel. Although comparing the volumes of the two gases may be like comparing apples with oranges, it is important to have these figures in mind (SO_2 produced by the truck fleet is essentially negligible).[rr]

Of course, not all of the 45,000 tonnes of cargo may want to shift to road. The proportion that will do so will depend, among other things, on things such as:

(a) the unit fuel costs of each of the two options (both for low-sulphur and for high-sulphur fuel)
(b) how the road option is exercised (e.g., it could be 1,125 trucks doing one trip each, a fleet of 563 trucks doing two trips each, or any other combination)
(c) the transit times of each of the two options
(d) the inventory costs of the cargo.

Regarding (a), we note the differential of $0.33/tonne of cargo in the price of fuel (or $14,809 per shipload, or some 66% more). This will translate into a cost increase of the sea mode. The calculation of the impact of this cost increase on the modal split between sea and road (which also depends on points (b), (c) and (d)) was an issue that was open at the time of the writing of this paper.

9. Conclusions

This paper has taken a look at some of the problems associated with green maritime logistics. Speed reduction was the main focus of the paper and some conditions under which such a scheme would reduce overall cost were identified for some operational scenarios. In addition, some possible ramifications of using speed reduction and cleaner fuels at SECAs were investigated. It was seen that caution should be exercised in proposing measures that may at first glance look environmentally friendly, but in reality they may have negative side effects.

[rr]This analysis does not take into account the additional CO_2 emitted by refineries to produce increased amounts of low-sulphur fuel, or the additional CO_2 as a result of the possible congestion by having a large number of trucks on the highway.

The main conclusions of this paper can be summarized as follows:

- Speed reduction will always result in a lower fuel bill and lower emissions, even if the number of ships is increased to meet demand throughput.
- Due to in-transit cargo inventory costs and other ship costs, total fleet operational costs may or may not decrease with speed reduction, depending on the scenario.
- The cost to avert one tonne of CO2 by speed reduction depends on several factors, being higher for higher-value cargoes.
- Speed reduction to reduce sulphur emissions at SECAs will result in a net increase of total emissions (including sulphur) along a ship's route, if transit time is to be kept the same.
- Cleaner fuel at SECAs may result in a reverse cargo shift from sea to land that has the potential to produce more emissions on land than those saved at sea.

Future research vis-a-vis the models presented here involves extending these models to more complex logistical scenarios, concerning, among other things, issues such as ship routing and scheduling, maritime and intermodal transport network design, queuing at ports, emissions at ports, and all the others listed at the introductory section. Plus, the broader consideration of such issues in a strategic setting is also important. In the emerging drive for green maritime logistics, investigating such problems would become increasingly important in the future.

Acknowledgments

This is an expanded version of the plenary address by the same title, given by the first author at the International Symposium on Maritime Logistics and Supply Chain Systems, held in Singapore in April 2009 (MLOG, 2009). Parts of the paper draw from papers presented at the International Marine Design Conference (IMDC, 2009), held in Trondheim, Norway, in May 2009, and the Conference on FAST Sea Transportation (FAST, 2009), held in Athens, Greece, in October 2009. The research reported in this paper was funded in part by a gift from Det Norske Veritas (DNV) to the National Technical University of Athens. Some of the data used in the paper was provided by the Hellenic Chamber of Shipping and by DNV.

References

1. Andersson, L. (2008). Economies of scale with ultra large container vessels, MBA assignment no 3, The Blue MBA, Copenhagen Business School.

2. Buhaug, Ø., Corbett, J.J., Endresen, Ø., Eyring, V., Faber, J., Hanayama, S., Lee, D.S., Lee, D., Lindstad, H., Mjelde, A., Pålsson, C., Wanquing, W., Winebrake, J.J. and Yoshida, K. (2008). Updated Study on Greenhouse Gas Emissions from Ships: Phase I Report. International Maritime Organization (IMO) London, UK, 1 September, (included as Annex in IMO document MEPC58/INF.6).

3. Cerup-Simonsen, Bo. (2008). Effects of energy cost and environmental demands on future shipping markets, MBA assignment no 3, The Blue MBA, Copenhagen Business School.

4. Congressional Budget Office (CBO) (2006). The Economic Costs of Disruptions in Container Shipments, U.S. Congress, Washington, DC.

5. Corbett, J.J. and Köhler, H.W. (2003). Updated emissions from ocean shipping, Journal of Geophysical Research 108.

6. Corbett, J., Wang, H. and Winebrake, J. (2009). Impacts of Speed Reductions on Vessel-Based Emissions for International Shipping, Transportation Research Board Annual Meeting, Washington DC.

7. Cordeau, J., Gaudioso, M., Laporte, G. and Moccia, L. (2007). The service allocation problem at the Gioia Tauro maritime terminal, *European Journal of Operational Research*, 176, 1167–1184.

8. Eefsen, T. (2008). Container shipping: Speed, carbon emissions and supply chain, MBA assignment no. 2, The Blue MBA, Copenhagen Business School.

9. Endresen, Ø., Sørgard, E., Sundet, J.K., Dalsøren, S.B., Isaksen, I.S.A., Berglen, T.F. and Gravir, G. (2003). Emission from international sea transportation and environmental impact, *Journal of Geophysical Research*, 108.

10. Endresen, Ø., Sørgård, E., Behrens, H.L., Brett, P.O. and Isaksen, I.S.A. (2007). A historical reconstruction of ships fuel consumption and emissions, *Journal of Geophysical Research*, 112.

11. EMEP/CORINAIR (2002). EMEP Co-operative Programme for Monitoring and Evaluation of the Long Range Transmission of Air Pollutants in Europe, The Core Inventory of Air Emissions in Europe (CORINAIR), Atmospheric Emission Inventory Guidebook, 3rd edition.

12. IMO (2000). Study of Greenhouse Gas Emissions from Ships, Study by Marintek, *Econ Centre for Economic Analysis*, Carnegie Mellon University and DNV.

13. IMO (2008a). INTERIM GUIDELINES FOR VOLUNTARY SHIP CO2 EMISSION INDEXING FOR USE IN TRIALS, MEPC Circ. 471.

14. IMO (2008b). Report of the Drafting Group on amendments to MARPOL Annex VI and the NOx Technical Code, MEPC 58/WP.9.

15. IMO (2008). Liaison with the Secretariats of UNFCCC and IPCC concerning the Carbon to CO2 conversion factor, MEPC 58/4/3.

16. IMO (2008). The impact of sulphur limits on ferry operations in Northern Europe, INTERFERRY, MEPC 58/5/11.

17. Kim, K.H. and Park, Y.M. (2004). A crane scheduling method for port container terminals, *European Journal of Operational Research*, 156, 752–768.

18. Lee, L., Chew, E., Tan, K. and Han, Y. (2006). An optimization model for storage yard management in transshipment hubs, *OR Spectrum*, 28, 539–561.

19. Lee, Y. and Hsu, N.Y. (2007). An optimization model for the container premarshalling problem, *Computers and Operations Research*, 34, 3295–3313.

20. Lim, A., Rodrigues, B., Xiao, F. and Zhu, Y. (2004). Crane scheduling with spatial constraints, *Naval Research Logistics*, 51, 386–406.

21. Lloyds List. (2008a). IMO Sulphur Limits Deal Could See More Freight Hit the Road, Lloyds List, 10 April.

22. Lloyds List. (2008b). An Efficient Ship is a Green Ship, says GL, Lloyds List, 30 July.

23. Psaraftis, H.N. and Kontovas, C.A. (2008). Ship Emissions Study. National Technical University of Athens, report to Hellenic Chamber of Shipping, http://www.martrans.org/emis/emis.htm, May.

24. Psaraftis, H.N. and Kontovas, C.A. (2009a). CO_2 Emissions Statistics for the World Commercial Fleet, *WMU Journal of Maritime Affairs*, 8(1), pp. 1–25.

25. Psaraftis, H.N. and Kontovas, C.A. (2009b). Ship Emissions: Logistics and Other Tradeoffs. 10th Int. Marine Design Conference (IMDC 2009), Trondheim, Norway, 26–29 May.

26. Psaraftis, H.N., Kontovas, C.A. and Kakalis N. (2009c). Speed Reduction as an Emissions Reduction Measure for Fast Ships. 10th Int. Conference on Fast Transportation (FAST 2009), Athens,Greece, 5–8 October.

27. Steenken, D., Voss, S. and Stahlbock, R. (2004). Container terminal operation and operations research — a classification and literature review, *OR Spectrum*, 26, 3–49.

28. Vis, I. and de Koster, R. (2003). Transshipment of containers at a container terminal: An overview, *European Journal of Operational Research*, 147, 1–16.

29. Wang, F. and Lim, A. (2007). A stochastic beam search for the berth allocation problem, *Decision Support Systems*, 42: 2186–2196.

CHAPTER 13

EXPLORING TANKER MARKET ELASTICITY WITH RESPECT TO OIL PRODUCTION USING FORESIM

P.G. Zacharioudakis

OceanFinance Ltd, Greece

Pzacharioudakis@oceanfinance.gr

D.V. Lyridis

School of Naval Architecture & Marine Engineering

National Technical University of Athens, Greece

dsvlr@central.ntua.gr

Future market freight levels have always been a critical question in decision support processes. FORESIM is a simulation technique that models shipping markets (developed recently). In this paper we present the application of this technique in order to obtain useful information regarding future values of the tanker market in numerous states of OPEC oil production levels. This is the first attempt to express future tanker market freight levels in relation to current market fundamentals and future values of demand drivers. We follow a systems analysis seeking for internal and external parameters that affect market levels. Therefore we apply dynamic features in freight estimation taking into account all Tanker market characteristics and potential excitations from non systemic parameters as well as their contribution to freight level formulation and fluctuation. In this way we are able to measure the behavior of future market as long as twelve months ahead with very encouraging results. The output information is therefore useful in all aspects of risk analysis and decision making in shipping markets.

1. Introduction

A popular definition of "forecast" is that it is a reference to future trends usually in the form of probability that is realized by processing and analyzing available data. Then a set of questions slip into mind: In a volatile market such as the one of shipping freight rates, is it possible to acquire information regarding its future evolvement? How can we predict the events that will influence the future state of the market? Future shipping market risk has always been an attractive thematic issue for many maritime economists. Especially in the era of risk management attempts in shipping, measuring market risk is a key point to the success of the whole process. Selecting between Spot and Period Charter or where to place a vessel, is a tricky question which, nevertheless, can be successfully approached by using the appropriate risk management tools. There are two major characteristics of the shipping market that turned risk management to a necessity: variability and uncertainty. Risk management can do very little to reduce market variability, but can be very effective in reducing uncertainty for those involved in risk-taking decisions. Alternatively Freight Forward Aggrements (FFA's) are the latest tools in hedging shipping risk aiming to be a gowning market parallel to the physical one. So whether we speak about physical market chartering strategies or "paper" market trading strategies, future market risk knowledge is essential for efficient decision making.

There are two main approaches regarding the estimation of future market risk. The first one is based on univariate stochastic models. This approach applies models like the GARCH, ARIMA, Geometric Brownian Motion (GBM), Ornstein-Uhlenbeck (O-U) process, Jump-Diffusion (O-U with Jumps) etc based on the admission that all the necessary information to estimate future values is located in the precedent historical data of the time series. This admission is quite defective by a simple consideration of the tanker and bulk shipping market mechanism. For that reason researchers developed static econometric models (Zannetos, 1966, Norman, 1979 & 1981, Evans 1994) or dynamic (Eriksen & Norman, 1976, Strandenes, 1986, Beenstock & Vergotis, 1989, Lyridis *et al.* 2004a & b).

Although this paper presents an application in Tanker vessels and more specifically in Very Large Crude Oil Carriers (VLCC), the methodology can also be applied in the bulk market. Both markets operate in a system with numerous interactions. Shipping market mechanism is full of causality terms balancing the output in the time field — the freight rates. The formulation mechanism for the tanker market is not clearly known but the fact is that it is dependent on the global socioeconomic status. The interaction of the market and the variables is either direct or indirect according to the way and the time lag that they interact. For example VLCC rates have a direct and positive correlation with the orderbook in real time. As the observed phenomenon of the Shipping cycle describes, following a high freight rate period new vessels enter the market resulting into an increase in the total transport capacity and subsequently into a drop in the rates. Therefore, the two variables have a negative correlation when examined under a specific time lag. All variables, apart from demand for sea transport, are in some way correlated to market trends and vice versa. However, demand for sea transport is determined by other factors and not by the state of the shipping market. For example, while the level of oil production by OPEC has a strong influence in the market, there is no feedback from the shipping market to the level of oil production. But which are the variables that influence demand in the shipping market? In the case of VLCC carriers the demand is related mainly to the following:

- The growth of world economy
- Oil shocks
- War — hostile acts near oil production facilities
- Oil reserves
- Oil price
- Climate conditions
- Political decisions — OPEC policy
- New reserves

Many shipping parameters have a dominant role in freight rates future possible realizations. This leads to the statement that the initial state of the shipping system as described by the fundamental variables has a leading effect on how this system may react to excitations such as a demand increase or decrease, a pipeline closure, an oil shock etc. To be more specific, a congested shipping market with increased volumes of laid up vessels is expected to show less sensitivity to demand changes comparing to a balanced market. It is known high volumes of laid up vessels is a characteristic of markets with low freight rates. The laid up vessels will absorb any demand increase by entering to operational state. To the contrary scenario if the market experiences a demand decrease for transport services then the already low freight rates levels cannot fall down from a minimum point relatively to the operating expenses. Another crucial parameter is the volume of tonnage under construction (order book). This parameter seems to give an important indication for the future levels of tonnage supply. Hence it is obvious that a systemic modeling of shipping market would lead to more bounded possible future states subject to the constrains of the fundamental explanatory shipping parameters. Forecasting the need for sea transport is very difficult since it is related to quantitative and qualitative variables with unforeseen trends. What can be done is to "feed" the forecasting model with different scenarios and generate a stochastic 'description' of the future. This is why FORESIM was conceived as a complete simulating procedure. The entire shipping market parameters such as active fleet or scheduled deliveries play a predetermined role in future freight rate levels. Additionally crucial parameters that affect freight rate levels (the OPEC oil production in our case) and have unpredictable random behaviors are stochastically generated in the corresponding simulation time period.

This paper is structured as follows: methodology is described in the next section. FORESIM technique is applied in order to simulate tanker VLCC market. The third section presents the results regarding future market estimation. The paper finishes with interesting conclusions about FORESIM application and market characteristics.

2. Methodology

FORESIM is a simulation technique developed especially for the shipping system. It is used in order to obtain a solid future view of the maritime trends. It consists of a stochastic model: this simulates oil production levels and then using the Monte Carlo technique produces the freight rates by applying a proper Artificial Neural Network. A main feature of FORESIM technique and what forms its basic innovative aspect is its ability to simulate future evolution of an econometric system based on its current state only, in a way where all crucial parameters can affect future system outputs. Hence in our case the entire shipping market parameters such as active fleet or scheduled deliveries play a predetermined role in future freight rate levels. Additionally crucial parameters that affect freight rate levels (the OPEC oil production in our case) and have unpredictable random behaviors are stochastically generated in the corresponding forecast time period.

The first step is the definition of the system and the variables to be simulated. This is what is called a systemic analysis in order to obtain absolute knowledge of the examinant system. Next step is to construct a model capable of simulating the physical market. FORESIM uses the power of Artificial Neural Models in what is called function approximation seeking for relations between input vector and desired output. Artificial neural networks are mathematical models imitating human brain functionality

and are used as an advanced pattern recognition technique with application
in time series forecasting. According to the literature, ANNs are suitable
for analysis of non-stationary nonlinear time series. Focusing in tanker
freight forecasting, in comparison to other methods such as linearly based
autoregressive models, artificial neural networks are proven to be at least
as accurate while, in many cases, yielding impressive results (Lyridis
2004a & b). The possible outer system excitations are entered into the
model with the usage of GARCH-family models. Then to model processes
like oil production, where covariance is not constant in the time domain,
GARCH family models has been fairly successful (Bollerlev, and Engle
1994).

The systemic analysis showed that the independent variables that can
be divided in two major categories. The first has to do with variables related
to demand for transport in and the second with those that are related to
the supply of tonnage. Figure 1 shows a schematic approach of the shipping
system regarding the freight rates generation mechanism.

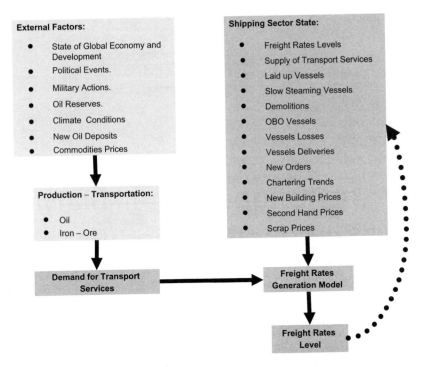

Fig. 1   Schematic approach of freight rates generation mechanism.

The importance of investigating these sets of variables is very high since they determine the freight rates as the result of the equilibrium between supply and demand. A few from the variables used in modeling the market are the following:

- Freight rates
- Active fleet
- Demand for transport in the specific market
- Orderbook
- Demolitions
- Laid-up vessels etc.

By using expert judgment and statistical tools to measure correlation and to avoid colinearity, the input vector is constructed. Table 1 shows the input vector for the three months ahead simulation of VLCC WS freight rates (Ras Tanura — Rotterdam) and the corresponding Variance Inflation Factor (VIF). The input vector consists of 8 variables:

By applying the same process the corresponding input vector for the twelve months ahead is shown in Table 2.

Table 1   Input vector for three months ahead model.

Independent variable	VIF
Oil price	2.858
VLCC supply	4.213
OBO supply	6.260
VLCC demolition prices	1.946
VLCC WS rate	3.727
OPEC production	4.148
OPECDIF_3 (percentage difference after three months)	1.104
ARBITRAGE of oil prices	4.941

Table 2   Input vector for twelve months ahead model.

Independent variable	VIF
VLCC supply	3.179
OBO supply	8.144
VLCC demolition prices	1.749
VLCC orderbook	5.001
WS	2.829
OPEC production	4.211
OPECDIF_12 (percentage difference after twelve months)	2.029

It is remarkable that as expected the Orderbook variable is shown to have a statistical important influence on the dependent variable of freight rates after twelve months. It also expected that variables like oil price and Arbitrage can influence the freight rate generation mechanism only in a short term basis. By constructing appropriate input vectors for up to a twelve months period ahead FORESIM is capable of simulating the VLCC market.

Oil production time series data like many time series usually exhibit a characteristic known as volatility clustering, in which large changes tend to follow large changes, and small changes tend to follow small changes. Engle's test is applied in Oil production time series data to seek for the presence of ARCH effects. Pre-estimation process includes the Opec time series transformation using the following equation:

$$Opecret_i = \frac{Opec_{i+1}}{Opec_i} - 1 \tag{1}$$

Under the assumption that the transformed Oil production time series data is a random sequence of Gaussian disturbances (i.e., no ARCH effects exist), this test statistic is also asymptotically Chi-Square distributed (Engle, 1995). The test results reveal that the ARCH effect is present hence serial dependence of volatility exists. Following this specific preprocessing procedure by applying Ljunx-Box-Pierce Q-test (Gourieroux, 1997) it is clear that no serial dependence of mean exists hence there is no need to use a conditional mean model such as ARIMA.

To feed the technique with possible future oil production volumes after a fit process an Exponential GARCH model is used. In order to fit a model in data set, log-likelihood function — LLF-criterion is calculated. In addition, Akaike and Bayesian information criteria were used to compare alternative GARCH models based on parsimony and penalize models with additional parameters. (Box, and Jenkins, 1970). Table 3 shows the results.

The E-GARCH(1,1) (Student t distributed) include a term to capture the leverage effect, or negative correlation, between examinant variable

Table 3   Input vector for twelve months ahead model.

Model	Selection Criteria		
	LLF	AIC	BIC
EGARCH11ARMA00T	819.07	−1628.15	−1608.34

Table 4   Input vector for twelve
months ahead model.

Coefficient	Value
C	0.00011918
K	−0.10279
GARCH(1)	0.98584
ARCH(1)	0.18242
Leverage(1)	−0.10488

returns and volatility (Nelson 1991). As estimated in the fit process the model will have the following coefficients as shown in Table 4.

Hence the form of the E-GARCH that will be used to generate paths of transformed Opec oil production is the following:

$$y_t = 0.00011918 + \varepsilon_t$$

$$\mathrm{Var}_{t-1}(y_t) = E_{t-1}(\varepsilon_t^2) = \sigma_t^2$$

$$\log \sigma_t^2 = -0.10279 + 0.98584 \log \sigma_{t-1}^2 \tag{2}$$

$$+ 0.18242 \left[\frac{|\varepsilon_{t-1}|}{\sigma_{t-1}} - E\left(\frac{|\varepsilon_{t-1}|}{\sigma_{t-1}} \right) \right] - 0.10488 \left(\frac{\varepsilon_{t-1}}{\sigma_{t-1}} \right)$$

Where:

$$E\{|z_{t-1}|\} = E\left(\frac{|\varepsilon_{t-1}|}{\sigma_{t-1}} \right) = \sqrt{\frac{\nu - 2}{\pi}} \cdot \frac{\Gamma\left(\frac{\nu-1}{2} \right)}{\Gamma\left(\frac{\nu}{2} \right)},$$

Due to the fact that ε_t is Student's T distributed.

The estimated ν freedom degrees equal to3.2314 for the specific distribution hence by calculating the Gamma function values, the term $E\left(\frac{|\varepsilon_{t-1}|}{\sigma_{t-1}} \right)$ takes the value of 0.6609.

By substituting the $E\left(\frac{|\varepsilon_{t-1}|}{\sigma_{t-1}} \right)$ term to the $\log \sigma_t^2$ equation:

$$\log \sigma_t^2 = -0.172109 + 0.98945 \log \sigma_{t-1}^2$$

$$+ 0.19252 \left(\frac{|\varepsilon_{t-1}|}{\sigma_{t-1}} \right) - 0.092892 \left(\frac{\varepsilon_{t-1}}{\sigma_{t-1}} \right) \tag{3}$$

A common question is which network to use in each case, as the researcher is faced with a large number of options. In this paper model the relation between the dependent and independent variables we use a special class of MultiLayer Perception networks (hereafter MLP), the modular feed-forward networks (Fig. 2).

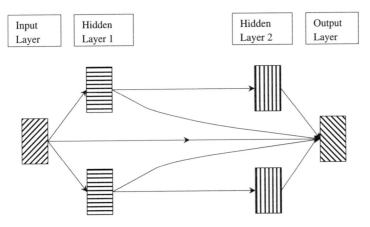

Fig. 2   Structure of the general modular artificial neural network.

These networks are trained by a supervised learning momentum algorithm (Moreira, 1995). The weight update process is the following:

$$w_{ij}(n+1) = w_{ij}(n) + \eta \cdot \delta_i(n)x_j(n) + a(w_{ij}(n) - w_{ij}(n-1)) \qquad (4)$$

Where:

η: learning rate
$\delta_i(n)$:   current error
$x_j(n)$:   current input vector
α: momentum rate

Modular ANN don't have full interconnectivity between neurons and the layers are divided to modules. Each module cooperates with others in order to solve part of the whole problem. Due to the partial interconnectivity a decreased number of weights is necessary and therefore the demand for training cases is decreased. The specific topology has two hidden layers with two modules per layer. The number of neuron per module is variable, subject to optimization. The transfer function is shown in Fig. 3:

$$\tanh(x) = (e^x - e^{-x})/(e^x + e^{-x}) \qquad (5)$$

And the output of the transfer is given by the following equation:

$$\text{output} = \tanh(w_1x_1 + w_2x_2 + \cdots + w_nx_n) \qquad (6)$$

It has to be mentioned that a modeler should consider basically two issues in order to obtain the ability of generalization (Jhee, M.J. Shaw). First, the explanatory model should transfer all necessary information to

tanh x

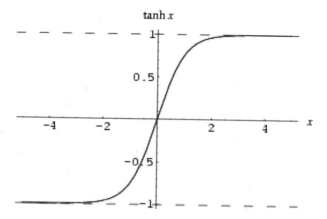

Fig. 3   Transfer function of the general modular artificial neural network.

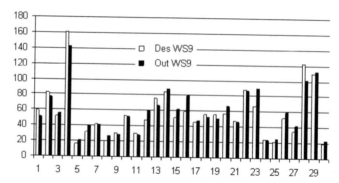

Fig. 4   Goodness transfer function of the general modular artificial neural network.

ANN. It is a matter of experience, deep knowledge and assiduous research effort to invoke all significant informational variables. The second crucial issue is to train the ANN using as a training data set a representative sample of data on which ANN will be used to simulate a forecast. ANN are trained using a cross validation dataset in order to avoid overtraining issues and luck of generalization ability. The cross validation data set consist of randomly selected cases which are kept out of the training process. By this way the ANN are trained and validated under a wider range of shipping market situations. Figure 4 shows example of ANN's fit on the corresponding test dataset (30 cases) for the nine months ahead case.

The results shows that fit on test data is excellent. This means that the information provided to the networks — current market fundamental variables and future demand indicator — is sufficient in order to estimate

Table 5   Goodness of fit results for ANN.

Goodness of fit criterion	ANN(+3)	ANN(+6)	ANN(+9)	ANN(+12)
MSE	0.003703	0.002843	0.006776	0.002872
NMSE	0.082971	0.081254	0.087043	0.075537
%Error	13.665811	11.780595	13.676121	14.865287

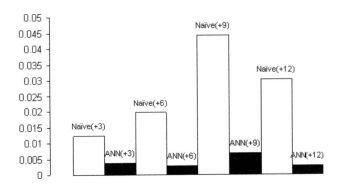

Fig. 5   Naïve model vs modular artificial neural network performance.

future market values. Table 5 shows the results of goodness of fit process using Mean Square Error, Normalized MSE and percentage of error criteria:

Comparing the results with the naïve model (future value is equal to the present one) it is clear that modular ANNs show a sufficient and robust error performance in the corresponding time spans of three, six, nine and twelve months. Figure 5 shows the results for the three, six, nine and twelve months ahead models:

The next step of FORESIM is simulation. The stochastic component (E-GARCH) feeds ANN with a pre defined number of possible future demands indicators and the corresponding input vector containing the fundamental variables describing the current state of the market. Then the ANN produce an output vector for the corresponding time span. This vector represents the estimated future freight rate values for every possible excitation from the demand indicator. The error terms of the ANN are stochastically estimated using a number of goodness of fit tests: chi-square (Snedecor and Cochran, 1989), Kolmogorov-Smirnov (Chakravart, Laha, &Roy, 1967) and Anderson-Darling (Stephens, 1974). By using simple Monte Carlo simulation every output value of the ANN is recalculated by adding a possible error term. An example of a FORESIM simulation case

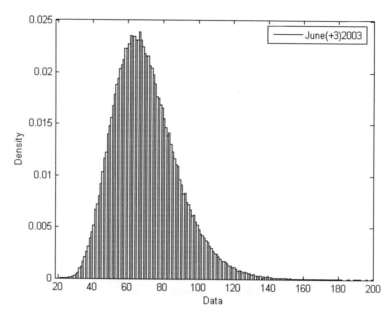

Fig. 6   Histogram of simulation results (case: three months ahead June 2003).

(simulated case: WS rate three months ahead June 2003) can be shown in
the next histogram (Fig. 6).

3. Simulation Results

The superior advantage of FORESIM technique is the ability of simulating
future market states in accordance to the fundamentals parameters and
possible external excitations. According to FORESIM results a crucial
parameter regarding future market levels is the level of OPEC oil production. By estimating the output of various cases of future oil production
an investigation regarding the relationship between future market and oil
production is feasible. Table 6 shows the bivariate Pearson Correlation
Coefficient between Demand for VLCC Tankers and the variable of Opec
oil production.

Table 6   Correlation of demand with opec production.

Pearson correlation	Opec production	Demand for tanker services
Opec production	1	.918
Demand for tanker services	.918	1

Table 7   Correlation of market rates with opec production.

Pearson correlation	Opec production	VLCC WS
Opec Production	1.000	.641
VLCC WS	.641	1.000

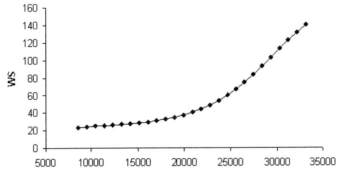

Fig. 7   WS vs OPEC oil production (millionbarrels per day) (12 Months Ahead March 2003).

Table 7 shows the bivariate Pearson Correlation Coefficient between VLCC Tankers freight rates and the variable of Opec oil production.

The results depict the importance of Oil production in future market freight levels as shown from Correlation Coefficients values. During a What — If analysis the output of a system is examined after considering various excitations. Shimojo (1979) has proposed a "J-shaped" diagram to describe freight rates versus demand for transportation. FORESIM is capable of simulating the future freight rates levels of a specific shipping system by applying various future oil production values. By assuming a specific starting point with well defined supply attributes, Orderbook value, oil production etc the method is capable of estimating the relation between future freight rates and future OPEC oil production. For example in March 2003 oil production was 28684 million barrels per day and the freight rate at Ras Tanura — Rotterdam route was 111 WS. Assuming Ceteris paribus for all shipping variables Fig. 7 shows the relation between future (after 12 months) freight rates and future (after 12 months) oil production:

- Due to very high values of orderbook at that time (47.8 mil ton DWT), the large delivery rate would push freight rates to lower levels if demand for transport services would stay constant. In fact after 12 months although oil production increased by 2% freight rates decreased from 111 to 98 WS. According to Fig. 7 in order freight market to hold at 111WS the oil production had to increase at 30300 million barrels per day.

- According to the j-shape curve for oil production below 20000 million barrels per day the balance is lower freight rates but as oil production increase the slope of the curve also increase.
- Although ANN have the ability to generalize close to the training values, for oil production values over the 40000 million barrels per day and lower than 15000 million barrels per day there is an uncertainty regarding the results due to the fact that these extreme values were not in the training set (outliers).

Figure 8 shows simultaneously the estimations for 12 months after January 2003 (red curve) and March 2003 (blue curve). The shipping fundamentals are more or less the same hence the expected reaction of the system is similar.

By applying totally different shipping system fundamentals to the method the results are totally different. At June 1994 the percentage of laid up vessels was 34.4% while at March 2003 it was only 6%. The fundamentals at June 1994 show a non-volatile shipping system to oil production fluctuation. This is shown at Fig. 9.

Figure 10 shows the reaction of the shipping system at two totally different time periods. At May 1987 there is a large amount of laid up vessels and the OPEC oil production is at 18000 million barrels per day when at June 1994 it was 26000 million barrels per day.

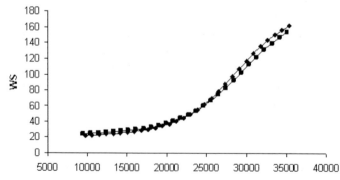

Fig. 8   WS vs OPEC oil production (millionbarrels per day) (12 months ahead March 2003 — squares — and January 2003 — diamonds).

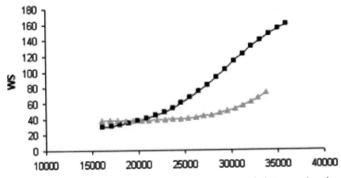

Fig. 9   WS vs OPEC oil production (millionbarrels per day) (12 months ahead March 2003 — squares — and June 1994 — diamonds).

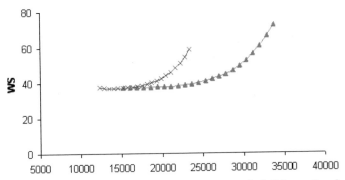

Fig. 10   WS vs OPEC oil production (millionbarrels per day) (12 months ahead May 1987 — crosses — and June 1994 — triangles).

As shown in Fig. 10 the curve of estimated freight rates after 12 months of May 1987 is shifted to the left compared with the one of June 1994. That complies with the fact that the available tonnage at May 1987 was 112 million ton. DWT while at June 1994 it was 130 million ton. DWT.

4. Conclusions

In this paper we presented the use of an innovative simulation technique. Although FORESIM was developed based on special shipping market characteristics, it has a wide range of applications in econometric systems. Focusing on shipping needs we concluded that risk is dominant in every decision. Our research aims to measure risk and provide initially to tanker owners a decisional framework to manage risk. To achieve this, firstly a tanker market analysis since 1979 was necessary in order to reveal shipping market mechanism, establish the most important factors affecting the market and decide whether a stochastic calculus was needed. The analysis showed that the crucial non shipping external variable affecting tanker market is the OPEC oil production level. This variable embodies political, economical, direct and indirect excitations to the shipping market. There is no better way to model and quantify excitations such as wars involving oil producing countries affecting productivity or economical crashes or OPEC decisions leading in many cases to oil shocks. Oil Production time series includes all economic and political facts in a global level that may affect a globalize market such as Tanker shipping market.

Subsequently, and keeping the aforementioned statement in mind, we tried to simulate the behavior of this variable using an E- GARCH model. When this was satisfactorily achieved, modular Artificial Neural Networks was trained to forecast future values of freight rates. Having real historical data for the route Ras Tanura –Rotterdam we constructed the ANN so as to predict the Tanker market after, three, six, nine and twelve month periods and tested it against randomly selected out of training sample data. To be precise, in fact, separate ANNs for each point in future were constructed, trained, and tested. The results showed that all ANNs were adequately capable of simulating future freight rates.

The special characteristics of FORESIM technique are shown in Table 8.

Table 8   Simulation vs Physical System.

	Systems	
	Physical shipping market	Simulated system with the use of FORESIM
Systems characteristics	Freight rate generation mechanism	Use of explanatory ANN to capture causality relations and interactions
	Random excitations from external non systemic parameters	Use of stochastic models to express randomness
	Non stationary system	Ability to add/remove parameters and readjust weights (adaptive system)
	Dynamic system-variability of shipping states in time	Ability to use technique in real time (real time output)

The procedure was developed in order to produce future freight rates realizations depended to the current state of the market. The procedure is the first to introduce the concept of generating freight rate realizations conditional upon the current or the preceding market states and of embedding explanatory and stochastic modeling. Therefore, it creates tool for acquiring quality information regarding the trend of the market taking into consideration unforeseen parameters as well as the present status of the market.

The main applications of FORESIM are the following:

- Decision support for trading Future Freight Agreements (FFAs) and various shipping market derivatives;
- Chartering strategy — spot or time charter, duration, etc.;
- Risk management for shipping investments, in combination with cash flow and monte carlo simulations providing distribution for financial variables;
- Estimation of financing risk such as probability of default etc.

In general it can be concluded that FORESIM represents one very promising tool in simulating freight rate time series.

References

1. Veenstra, A.W. (1999). Quantitative Analysis of Shipping Markets (Delft University Press Postbus).
2. Gourieroux, B.V.C. (1997). ARCH Models and Financial Applications, Springer-Verlag.
3. Beenstock, M. and Vergottis, A. (1989). An econometric model of the world tanker markets. *Journal of Transport Economics and Policy*, 23, pp. 263–280.
4. Beenstock, M. and Vergottis, A. (1993). Econometric Modeling of World Shipping (Chapman & Hall).
5. Berndt, E.R., Hall, B.H., Hall, R.E. and Haussman, J.A. (1974). Estimation and inference in nonlinear structural models, *Annals of Economic and Social Measurement*, 4, pp. 653–665.
6. Bollerlev, T. and Engle, R.F. (1994). Arch Models, Handbook of Econometrics, Volume IV, Elsevier Science.
7. Box, G.E. and Jenkins, G.M. (1970). Time Series Analysis: Forecasting and Control, Holden Day, San Francisco.
8. Drewry Shipping Consultants Ltd.: The Drewry Monthly.
9. Frees, E. (1996). Data Analysis Using Regression Models (Prentice Hall).
10. Enders, W. (1995). Applied Econometric Time Series (John Wiley & Sons).
11. Engle, R.F. (1995). ARCH Selected Readings, Advanced Texts in Econometrics (Oxford University Press).

12. Eriksen, I.E. (1977). The Demand for Bulk Services. Norwegian Maritime Research, 2, Págs. 22–26.

13. Evans, J.J. (1994). An analysis of efficiency of the bulk shipping markets, *Maritime Policy and Management*, 21(4), pp. 311–329.

14. Demuth, H. and Beale, M. (1996). Neural Network Toolbox for use with MATLAB (The Math Works Inc.).

15. Hentschel, L. (1995). All in the Family: Nesting Symmetric and Asymmetric GARCH Models, *Journal of Financial Economics*, 39, pp. 71–104.

16. Li, J. and Parsons, M. (1997). Forecasting Tanker Freight Rate Using Neural Networks (Maritime Policy and Management January Issue).

17. Jang, J.S.R., Sun, C.T. and Mizutani, E. (1997). Neuro-Fuzzy and Soft Computing (Prentice Hall).

18. Koopmans, T.C. (1939). Tanker Freight Rates and Tankship Building, Haarlem, The Netherlands.

19. Lloyd's: Shipping Economist, Clare Longley (publisher)

20. Lyridis D.V., Zacharioudakis P.G., Mitrou, P. and Mylonas, A. (2004a). Forecasting tanker market using artificial neural networks, *Maritime Economics and Logistics*, 6(2), pp. 93–108.

21. Lyridis, D.V., Zacharioudakis, P.G. and Ousantzopoulos, G. (2004b). A Multi-Regression approach to forecasting freight rates in the dry bulk shipping market using neural networks, Proceeding of the Annual Conference of the International Association of Maritime Economists, IAME 2004, Izmir, Turkey, Vol. II, pp. 797–811.

22. Azoff, M. (1994). Neural Network Time Series Forecasting of Financial Markets (Wiley).

23. Beenstock, M. and Vergottis, A. (1993). Econometric Modelling of World Shipping (Chapman & Hall).

24. Monthly Energy Review of the Energy Information Administration, U.S. Department of Energy (http://www.eia.doe.gov/).

25. Nelson, D.B. (1991). Conditional Heteroskedasticity in asset returns: A new approach, *Econometrica*, 59, 1991.

26. Norman, V.D. and Wergeland, T. (1981). Nortank — A simulation model of the freight market for large tankers. Centre for Applied Research, Norwegian School of Economics and Business Administration, Report 4/81. Bergen, Norway.

27. Norman, Victor D. (1979). Economics of Bulk Shipping. Institute for Shipping Research, Norwegian School of Economics and Business Administration, Bergen.

28. Pindyck R. and Rubenfeld, D. (1998). Econometric Models and Economic Forecasts (McGraw-Hill).

29. Trippi, R.R. and Turban, E. (1996). Neural Networks in Finance and Investing (Irwin).

30. Strandenes, S.P. (1986). NORSHIP — a simulation model for bulk shipping markets. Center for Applied Research, Norwegian School of Economics and Business Administration, Bergen, Norway.

31. Svendsen, A.S. (1958). Sea Transport and Shipping Economics, Bremen.

32. Crooks, T. (1992). Care and Feeding of Neural Networks.
33. Shimojo, T. (1979). Economic Analysis of Shipping Freights (Kobe University).
34. Timbergen, J. (1939). Selected Papers, Klaasen, L.H. *et al.* Editors, North Holland Publishing Company, Amsterdam, The Netherlands.
35. Timebergen, J. (1931). Ein Schiffbauzyclus? Weltwirtschaftliches Archiv, 34, pp. 152–64.
36. Jhee, W.C. and Shaw, M.J. Time Series Prediction Using Minimally Structured Neural Networks: An Empirical Test.
37. Zannetos, Z. (1966). The Theory of Oil Tankship Rates (MIT press).